Synthesis and modification of

bicyclo[1.1.1]pentyl sulfides

Zur Erlangung des akademischen Grades eines

DOKTORS DER NATURWISSENSCHAFTEN

(Dr. rer. nat.)

von der KIT-Fakultät für Chemie und Biowissenschaften

des Karlsruher Instituts für Technologie (KIT)

genehmigte

DISSERTATION

von

Robin Maximilian Bär

aus

Pforzheim

1. Referent: Prof. Dr. Stefan Bräse
2. Referent: Prof. Dr. Joachim Podlech
Tag der mündlichen Prüfung: 05.02.2020

Band 89
Beiträge zur organischen Synthese
Hrsg.: Stefan Bräse

Prof. Dr. Stefan Bräse
Institut für Organische Chemie
Karlsruher Institut für Technologie (KIT)
Fritz-Haber-Weg 6
D-76131 Karlsruhe

Bibliographic information published by the Deutsche Nationalbibliothek

The Deutsche Nationalbibliothek lists this publication in the Deutsche Nationalbibliografie; detailed bibliographic data are available in the Internet at http://dnb.d-nb.de

ISBN 978-3-8325-5109-4
ISSN 1862-5681

Logos Verlag Berlin GmbH
Comeniushof, Gubener Str. 47,
10243 Berlin
Tel.: +49 030 42 85 10 90
Fax: +49 030 42 85 10 92
INTERNET: http://www.logos-verlag.de

Organic chemistry is the child of medicine, and however far it may go on its way,

with its most important achievements, it always returns to its parent.

J.L.W. Thudichum

The present work was carried out at the Institute of Organic Chemistry at the Karlsruhe Institute of Technology (KIT), at Boehringer Ingelheim Pharma GmbH & Co. KG in Biberach an der Riß and at the School of Chemistry at the University of Bristol (UK) in the period from 1st January 2017 to 7th January 2020 under supervision of Prof. Dr. Stefan Bräse. During the period from 1st April to 30th September 2019 the work was scientifically supervised by Prof. Varinder Kumar Aggarwal (University of Bristol).

Die vorliegende Arbeit wurde im Zeitraum vom 1. Januar 2017 bis 7. Januar 2020 am Institut für Organische Chemie des Karlsruher Instituts für Technologie (KIT), bei Boehringer Ingelheim Pharma GmbH & Co. KG in Biberach an der Riß und an der School of Chemistry der University of Bristol (UK) unter der Leitung von Prof. Dr. Stefan Bräse durchgeführt. Während des Zeitraums vom 1. April bis 30. September 2019 lag die wissenschaftliche Betreuung bei Prof. Varinder Kumar Aggarwal (University of Bristol).

Hereby I declare, that I completed the work independently, without any improper help and that all material published by others is cited properly. This thesis has not been submitted to any other university before.

Hiermit versichere ich, die vorliegende Arbeit selbständig verfasst und keine anderen als die angegebenen Quellen und Hilfsmittel verwendet sowie Zitate kenntlich gemacht zu haben. Die Dissertation wurde bisher an keiner anderen Hochschule oder Universität eingereicht.

German title of this thesis

Synthese und Modifikation von Bicyclo[1.1.1]pentylsulfiden

Table of Contents

1 Abstract

Bicyclo[1.1.1]pentanes (BCPs) are one of the non-conjugated rigid hydrocarbons (NRHs) that gained interest in material sciences and as a non-classical bioisostere for *para*-substituted benzenes, alkynes and *tert*-butyl groups in drug design. The synthesis of BCPs is still challenging and limits their application. Starting from the strained [1.1.1]propellane (**1**) there have been many contributions using CC and CN bond formations to obtain BCPs. The CS bond formation has been rarely used and not systematically investigated.

Therefore, this thesis aimed at the development of methods to obtain BCP sulfides and related structures from **1**. The radical addition of thiols to **1** was found to be a versatile and easy method to prepare terminal BCP sulfides. The mild conditions tolerated a variety of functional groups and allowed the synthesis of 31 examples including a novel amino acid and a peptoid building block.

To obtain 1,3-disubstituted BCPs, disulfides were reacted with **1** under UV irradiation. This method allowed the selective synthesis of the BCPs instead of a mixture of BCPs and oligomers (staffanes) for the first time. Through variation in the ratio of the starting materials the product mixture could be influenced and some [2]staffanes were synthesized as well. The developed method was also applied to synthesize non-symmetrically substituted BCP sulfides from two symmetrical disulfides.

The obtained BCP sulfides were oxidized and iminated to obtain BCP sulfoxides and sulfoximines. These modifications to tune parameters like the polarity are important for a successful application of the structural motif in drug design. Together with the master student Lukas Langer a variety of BCP sulfoximines was synthesized and *N*-arylated using aryl halides and copper(I)-catalysis. Further, BCP sulfone could be applied as a directing group for an *ortho*-lithiation similar to *tert*-butyl sulfone.

The final aim of this thesis was the synthesis and application of a bench-stable BCP building block to facilitate the use in medicinal chemistry and other fields. Sodium BCP sulfinate was obtained in a four-step procedure in good yield starting from commercially available precursors without the need for chromatography or crystallization. The sulfinate was applied in the synthesis of BCP sulfones, sulfoxides, a sulfinamide and sulfonamides. A desulfinylative cross-coupling of the salt using electrochemical oxidation is still under investigation in a collaborative project with Dr. Kevin Lam (University of Greenwich, UK).

2 Kurzzusammenfassung

Bicyclo[1.1.1]pentane (BCPs) gehören zu den nicht-konjugierten rigiden Kohlenwasserstoffen (engl. *non-conjugated rigid hydrocarbons*, NRHs), die in Materialwissenschaften und als nicht-klassische Bioisostere für *para*-substituierte Benzole, Alkine und *tert*-Butyl Gruppen im Wirkstoffdesign auf Interesse stießen. Die Synthese von BCPs ist immer noch eine Herausforderung und schränkt ihre Anwendung ein. Ausgehend von dem gespannten [1.1.1]Propellan (**1**) wurden zahlreiche Beiträge unter Verwendung von CC- und CN-Bindungsbildungen geleistet, um BCPs zu erhalten. Die Bildung von CS-Bindungen wurde selten genutzt und nicht systematisch untersucht.

Daher war das Ziel dieser Arbeit die Entwicklung von Methoden zur Darstellung von BCP-Sulfiden und Derivaten ausgehend von **1**. Die radikalische Addition von Thiolen an **1** erwies sich als vielseitige und einfache Methode zur Synthese terminaler BCP-Sulfide. Die milden Bedingungen tolerierten eine Vielzahl an funktionellen Gruppen und ermöglichten die Synthese von 31 Beispielen, einschließlich einer neuen Aminosäure und eines Peptoid-Bausteins.

Um 1,3-disubstituierte BCPs zu erhalten, wurden Disulfide unter UV-Bestrahlung mit **1** umgesetzt. Diese Methode ermöglichte erstmals die selektive Synthese von BCPs anstelle eines Gemisches aus BCPs und Oligomeren (Staffanen). Durch Variation des Verhältnisses der Edukte konnte das Produktgemisch beeinflusst und auch einige [2]Staffane synthetisiert werden. Die entwickelte Methode wurde auch angewendet, um unsymmetrisch substituierte BCP-Sulfide aus zwei symmetrischen Disulfiden zu synthetisieren.

Die erhaltenen BCP-Sulfide wurden oxidiert und iminiert, um BCP-Sulfoxide und Sulfoximine zu erhalten. Diese Modifikationen von Eigenschaften, wie z.B. Polarität, sind wichtig für eine erfolgreiche Anwendung des Strukturmotivs im Wirkstoffdesign. Zusammen mit dem Masterstudenten Lukas Langer wurde eine Vielzahl von BCP-Sulfoximinen synthetisiert und mit Arylhalogeniden und Kupfer(I)-Katalyse *N*-aryliert. Weiterhin konnte BCP-Sulfon als dirigierende Gruppe für eine *ortho*-Lithiierung ähnlich wie *tert*-Butylsulfon angewendet werden.

Das letzte Ziel dieser Arbeit war die Synthese und Anwendung eines stabilen BCP-Bausteins, um den Einsatz in der medizinischen Chemie und auf anderen Gebieten zu erleichtern. Natrium BCP-Sulfinat wurde in einer vierstufigen Synthese in guter Ausbeute ausgehend von kommerziell erhältlichen Vorläufern erhalten, ohne dass eine Chromatographie oder Kristallisation erforderlich war. Das Sulfinat wurde in der Synthese von BCP-Sulfonen, Sulfoxiden, einem Sulfinamid und Sulfonamiden angewendet. Eine desulfinyliernde Kreuzkupplung des Salzes mittels elektrochemischer Oxidation wird derzeit in einer Kollaboration mit Dr. Kevin Lam (University of Greenwich, UK) untersucht.

3   Introduction

The chemical space of small carbon-based molecules, with a molecular weight up to 500 g/mol, is estimated to contain more than 10^{60} compounds.[1] Even the most ambitious teams of synthetic chemists could only synthesize a fraction of this number. Besides, it would neither be efficient nor expedient to try. Biologically relevant chemical space covers only a small part of the chemical space and even less compounds are interesting for pharmacological applications (Figure 1).[2]

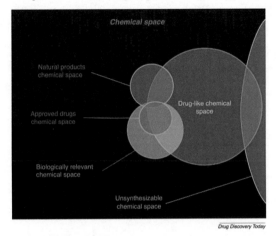

Figure 1. Chemical space visualized in two dimensions. Biologically relevant chemical space only covers a fraction of it and not all biologically relevant compounds fulfill the requirements for a modern drug. This figure was published in [3], © Elsevier (2018).

There are different approaches to guide medicinal chemists through chemical space towards a lead structure.[4] Random walks are rarely rewarded with success and a more directed approach is the diversity-oriented synthesis.[5-6] Alternatively, a known drug or a bioactive natural product can serve as a model.[4] In the target-oriented synthesis the biological target, e.g. an enzyme, has to be identified first and subsequently a therapeutic compound to influence this target needs to be developed, e.g. through fragment-based drug design.[7]

In all of the mentioned strategies, the availability of synthetic methods can limit the direction of research.[8-10] Therefore, the development of novel methods is an important factor towards new, more effective and safer drugs. In the following chapters the concept of bioisosterism as a technique in drug discovery, the use of bioconjugation to modify biomolecules and the history of [1.1.1]propellane (**1**) (and bicyclo[1.1.1]pentanes) will be discussed. The latter might offer applications in both areas, bioisosterism and bioconjugates.

3.1   Bioisosterism

Drug candidates need to fulfill a variety of requirements like potency, bioavailability, metabolic stability and many others. This makes drug discovery a multi-objective problem (MOP, Figure 2).[11] One tool to tackle this MOP is bioisosteric replacement. Thereby, certain atoms or groups of the lead structure are replaced by groups with similar shape and function.[12] The shape and function of a group in a biological environment is dependent on factors like molecular weight, bond angles, dipoles, pK_a, hydrogen bonding capacity and others. This broad range of properties makes it difficult to measure or exactly define bioisosterism and some consider the concept to be qualitative and intuitive.[13]

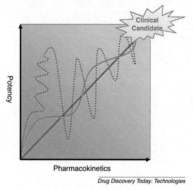

Figure 2. Drug discovery as multi-objective problem (MOP) exemplified with potency and pharmacokinetics as two factors. The iterative optimization of single properties (dashed line) requires more time and resources than the multi-objective optimization (orange continues line). This figure was published in [11], © Elsevier (2013).

Bioisosteres can be divided into classical and nonclassical bioisosteres.[14] Some examples for both classes will be discussed in this chapter to give an impression of the power and limitations of bioisosterism. Most examples were chosen based on a review by Patani and LaVoie.[15]

3.1.1   Classical bioisosteres

The classical bioisosteres are restricted to Grimm's Hydride Displacement Law[16-17] and the definition of isosteres by Erlenmeyer.[18] In short, atoms, ions and molecules in which the peripheral layer of electrons can be considered identical are defined as isosteres. Together with a similar biological activity, these restrictions define classical bioisosteres, which can be divided into five categories (Table 1).[14, 19]

Monovalent atoms or groups include, among others, hydroxyl groups, thiols and halides. The replacement of a hydrogen atom in a drug candidate with fluorine can have significant impact on the metabolic stability, as it can block the oxidation by Cytochrome P450 monooxygenases and therefore prevent a faster clearance. This effect was used during the lead optimization of Ezetimibe (**3**), a cholesterol absorption inhibitor, with a 50 fold lower effective dose (ED_{50}) than the initial compound **2** (Scheme 1 A).[20] With a similar van der Waal's radius of 1.35 Å (1.2 Å for hydrogen) fluorine does not change the steric demands of the substituent, which leads to almost the same potency in many cases.[19]

Table 1. Categories and some examples of classical bioisosteres. Groups in each row are equivalent.[12, 19]

Category	Example
A) monovalent atoms or groups	F, H OH, NH F, OH, NH, or CH_3 for H SH, OH Cl, Br, CF_3
B) divalent atoms or groups	$-C=S, -C=O, -C=NH, -C=C-$
C) trivalent atoms or groups	$-\overset{\displaystyle}{\underset{H}{C}}{=}, -N{=}$ $-P{=}, -As{=}$
D) tetrasubstituted atoms	$-\overset{\mid}{\underset{\mid}{N}}{}^{+}- \quad -\overset{\mid}{\underset{\mid}{C}}- \quad -\overset{\mid}{\underset{\mid}{P}}{}^{+}- \quad -\overset{\mid}{\underset{\mid}{As}}{}^{+}-$
E) ring equivalents	

Divalent atoms or groups can often be replaced without losing the activity of the compound. The example shown in Scheme 1 B comprises a guanosine analogue tested *in vivo* as antiviral agent.[21] Therefore, mice were infected with Semliki Forest virus (SFV), a virus that infects cells of the central nervous system, and treated with guanosine analogues **4**, which only differed in the heteroatom in position 8 (X). All three compounds showed antiviral activity and more than half the animals survived. In the placebo group only 1/11 mice (9%) survived the virus after 21 days. Similar replacements are possible for trivalent atoms or groups and tetrasubstituted atoms (see Table 1).

A typical example for bioisosteric ring equivalents is the replacement of the γ-lactone in the cholinergic muscarinic agonist Pilocapine (**5**, Scheme 1 C).[22] The natural product used to reduce intraocular pressure is only active for three hours due to lactone hydrolysis and subsequent elimination and/or epimerization.[23] By replacing the lactone with a cyclic carbamate compound **6** was designed and showed the same potency with a higher stability against hydrolysis.[22]

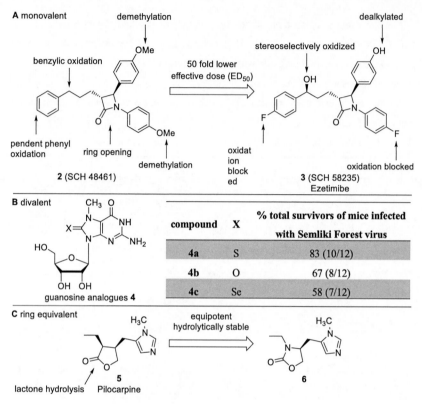

Scheme 1. Examples of classical bioisosteric replacement of A) monovalent atoms or groups,[20] B) divalent atoms or groups[21] and C) ring equivalents.[22]

3.1.2 Nonclassical bioisosteres

If a similar biological activity can be maintained, but the definitions for isosteres are not fulfilled, the term nonclassical bioisostere can be applied. In these cases, electronic, spatial or physicochemical properties that are critical for the biological activity are mimicked.[15]

For several functional groups nonclassical bioisosteres were established. The alkylsulfonamido group for example, can replace a phenolic group as the exchangeable proton possesses a similar acidity (Scheme 2 A).[24]

Another sulfonamide and the first systematically used antimicrobial drug Prontosil (**9**) introduced by Bayer in the 1930s[25] was later found to be a prodrug (Scheme 2 B).[26] The active form, Sulfanilamide (**10**), competes with *para*-aminobenzoic acid (PABA, **11**) in the bacterial biosynthesis of folate[27] and therefore can be classified as nonclassical bioisostere of **11**. Other bioisosteres of carboxylic acids include tetrazoles, sulfonates and phosphates.[15]

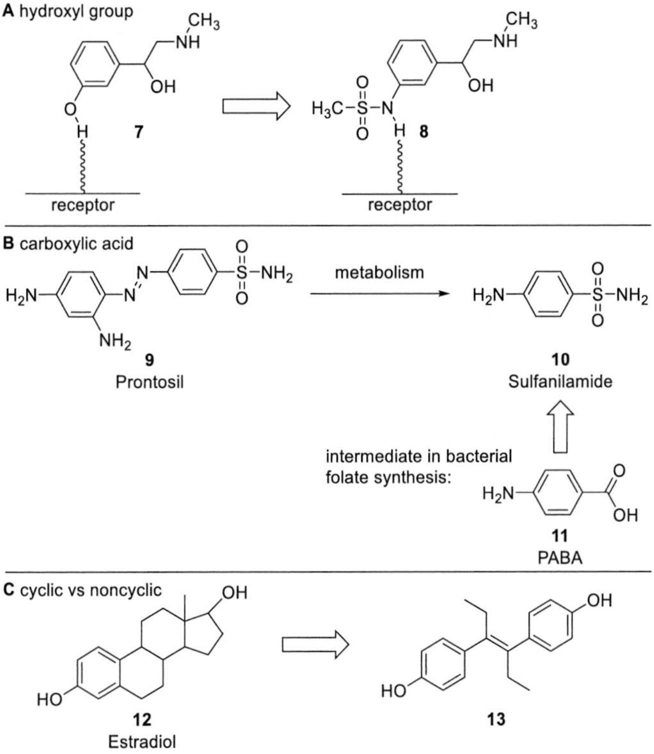

Scheme 2. Examples for nonclassical bioisosteres. A) Exchange of a phenol group with an alkylsulfonamide;[24] B) Sulfanilamide (**10**) as bioisostere to *para*-aminobenzoic acid (PABA, **11**);[27] C) replacement of a cyclic scaffold by a *trans* alkene.[28]

There are many bioisosteric replacements for amides, e.g. thioamides, ureas or esters, which will not be discussed in this thesis. Also, bioisosteres for halides will only be mentioned briefly. Most commonly halides can be replaced with cyano or trifluoromethyl groups. If the activity of the compound is dependent on the strength of the electron-withdrawing group, the replacement of halides with the mentioned groups is a handy tool to tune this parameter.[15]

A less obvious replacement is the exchange of a cyclic group for a noncyclic moiety (Scheme 2 C). In this case the rigidity of the noncyclic system is a key factor for a successful replacement. In the shown example *trans* alkene **13** could be used instead of the steroid hormone Estradiol (**12**) with similar potency.[28] The *cis* isomer of **13** and non-rigid analogues (not shown) did show little to none binding to the estrogenic receptor.[29-31]

3.1.3 Saturated hydrocarbons as bioisosteres

Among the nonclassical bioisosteres, saturated hydrocarbons (or non-conjugated rigid hydrocarbons, NRHs) represent a special subclass. Most commonly used as rigid linear linkers, they can replace *para*-substituted benzene rings or alkynes. Applications in material sciences will not be discussed in this thesis, but are nicely summarized in a recent review by Senge *et al.*[32]

NRHs play a central role in the concept 'Escape from Flatland' coined by Lovering *et al.*[33-34] They showed that an increasing complexity of a drug candidate, meaning a higher number of sp^3 hybridized carbons and the presence of a chiral carbon, correlates with a higher success rate in drug development. One reason for this is the limited chemical space accessed by high-throughput syntheses. As these syntheses usually are based on reliable couplings of available building blocks, they are mainly dominated by achiral, aromatic moieties. Another reason is that the exchange of aromatic structures with saturated hydrocarbons increases the solubility and bioavailability of drug candidates. Different scaffolds can be applied to tune the distance between two groups of the molecule (Figure 3). It's important to note, that a replacement of an aromatic structure with a NRH can only be successful, if no aryl-protein interaction is necessary for the activity of the compound.[35]

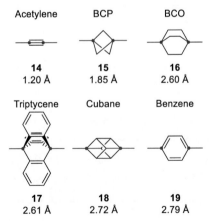

Figure 3. CC distances of different linear linkers. The NRHs (BCP, BCO, Triptycene, Cubane) can be used to tune the distance in a drug candidate. Figure based on Senge *et al.*[32]

An additional bioisosteric replacement should briefly be mentioned, *tert*-butyl groups can be replaced by terminal bicyclo[1.1.1]pentanes (BCPs), which sometimes leads to improved metabolic stabilities.[36]

The first example of a bioisosteric replacement of benzene with NRHs was reported by Pellicciari *et al.* (Scheme 3 A). The known antagonist of group I metabotropic glutamate receptors (mGluRs) **20** was compared to its bicyclo[1.1.1]pentane (BCP) and cubane analogs **21** and **22**.[37-38] While the BCP **21** showed an increased activity, the cubane **22** could not reach the potency of the model compound **20**. The authors concluded that the cubane spacer reached the upper limit for the steric accessibility of the receptor.[37]

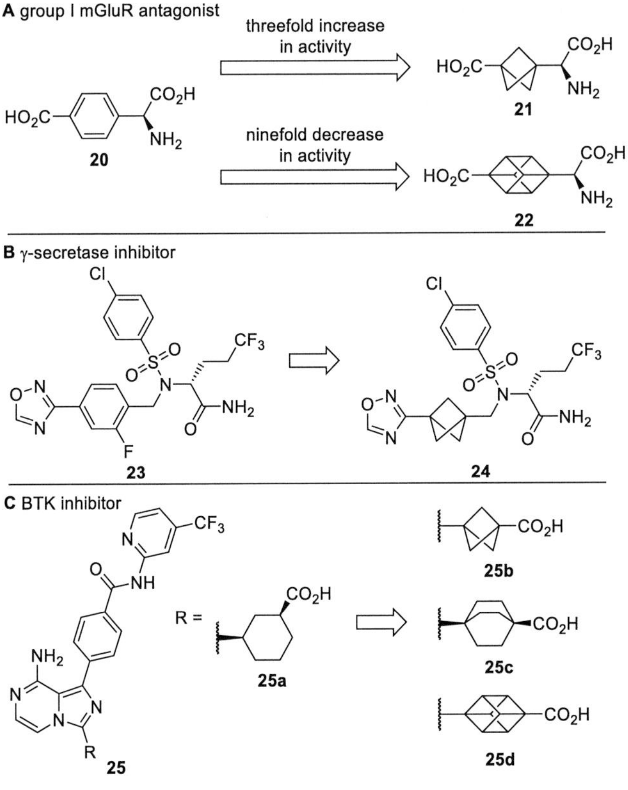

Scheme 3. Non-conjugated rigid hydrocarbons (NRHs) as bioisosteres in A) mGluR antagonists,[37-38] B) a γ-secretase inhibitor,[39] and C) BTK inhibitors.[40]

A milestone in the application of NRHs as bioisosteres was the study by Stepan *et al.* in 2012 (Scheme 3 B).[39] The γ-secretase inhibitor **23** was in clinical trials for the treatment of Alzheimer's disease.[41] Further investigations led to the synthesis of **24** with a BCP instead of the fluorinated benzene. The BCP **24** did not only show an increased potency (subnanomolar inhibition), but also better passive permeability, metabolic stability and solubility in water.[39] This discovery led to a significant increase in the use of BCP in drug candidates.[35] The fact that the new compound was patent-free without the benzene moiety was clearly another attractive feature of such replacements.

The Bruton's tyrosine kinase (BTK) inhibitors **25** were part of an extensive structure-activity relationship (SAR) study (Scheme 3 C).[40] Starting from the initial lead structure **25a** different linkers were used to vary the distance and the projection of the carboxylic acid. Among the numerous compounds BCP **25b**, bicyclo[2.2.2]octane (BCO) **25c** and cubane **25d** were tested. Although first results of **25c** were promising, the compound was not further investigated due to very high clearance.

The use of NRHs in drug discovery can often be challenging due to a lack of synthetic methods or building blocks.[32] Therefore, method development in this field is highly desired. The synthesis of BCPs will be in the focus of this work with a future application in bioisosteric replacements in mind.

3.2 Bioconjugation

Covalent bond formation between biomolecules or with synthetic compounds is called bioconjugation and this tool is omnipresent in biochemical research (Figure 4).[42] It's used to identify novel biomolecules, or targets related to certain diseases as well as in the investigation of complex biological processes on a molecular level. Also in medical applications like diagnostics and even in material sciences bioconjugates are applied. Only a few examples will be mentioned in this chapter, with a focus on bioconjugations with thiols and disulfides.

Figure 4. General principle of bioconjugation. With a conjugation agent two (bio-)molecules (A and B) can be connected. Figure based on Hermanson.[42]

There is a broad range of established techniques that can be categorized by the target biomolecule that should be addressed. The first and largest category includes amino acids, peptides and proteins. Less common targets are sugars, polysaccharides and glycoconjugates (second category), as well as nucleic acids and oligonucleotides (third category).[42]

When planning a bioconjugation experiment different aspects need to be addressed to decide which process is suitable. Stephanopoulos and Francis designed a flowchart that can be helpful during the planning of a protein bioconjugation (Figure 5).[43] An important distinction between site specific and nonspecific bioconjugation needs to be made. For nonspecific bioconjugations the reaction of *N*-hydroxysuccinimidyl (NHS) esters **27** with lysine (**26**) is probably the most prominent one (Scheme 4). As lysines are typically displayed on the outside of the protein (in aqueous solution) and virtually all proteins have many lysines, this reliable and general method is applied very often.[44] Many NHS esters are commercially available or easily prepared from the carboxylic acid.

A more specific bioconjugation can be achieved when cysteine is addressed, as cysteine is one of the rarest amino acids in native proteins.[45] Typical reactions of cysteine or the disulfide equivalent cystine will be discussed in the following chapter. More sophisticated methods like native chemical ligation and SNAP tagging will not be part of this thesis.

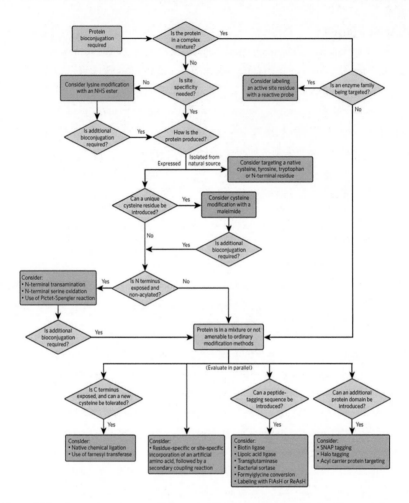

Figure 5. Flowchart to choose an appropriate protein bioconjugation strategy. Reprinted by permission from Macmillan Publishers Ltd: Nature Publishing Group, Nature Chemical Biology,[43] © 2011.

Scheme 4. Nonspecific bioconjugation of lysine (**26**) in proteins with NHS esters **27**.[44]

3.2.1 Conjugations with cysteine and cystine

The bioconjugation with cysteine (**29**) holds some advantages over reactions with other amino acids. The rareness in native proteins in particular makes cysteine a desirable target and it can be added to the sequence of modified proteins easily. The pK_a value of the cysteine side chain (~8) allows deprotonation in basic aqueous solutions. The generated thiolate is a strong nucleophile and can react with soft electrophiles like iodoacetamides **30** (Scheme 5 A) or maleimides **32** (Scheme 5 B). In cases of cross-reactivity of lysine with these electrophiles the pH value can be lowered to form the protonated amino group.[44]

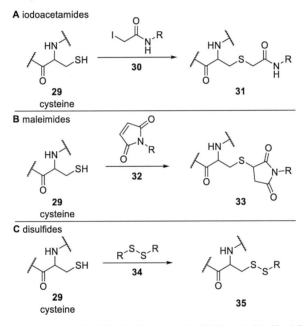

Scheme 5. Bioconjugations of cysteine (**29**) with A) iodoacetamides **30**; B) maleimides **32** and C) disulfides **34**.[44]

Another possible modification is the formation of disulfides **35** (Scheme 5 C).[42, 44] These reactions are often carried out with a large excess of the disulfide or with non-symmetrical disulfides with good leaving groups. Depending on the application of the conjugation the disulfide bond can be cleaved again, rebuilding the unmodified cysteine.

In many cases two native cysteines are oxidized to the disulfide (cystine), which can also stabilize tertiary structures in proteins. The reduction of the disulfides, e.g. with dithiothreitol or β-mercaptoethanol, enables the abovementioned methods for bioconjugation.[44] The two thiol groups can also be connected in a permanent way, e.g. through two alkylations, to stabilize the protein structure. This can lead to improved properties of proteins, hormones or antibodies for *in vivo* applications.[46] The distance of the rebridged disulfide can be tuned by using different alkylating agents.[47]

Bioconjugation with disulfides plays an important role in the development of antibody-drug conjugates (ADCs). With the aim of developing a potent drug, e.g. for cancer treatment with less side-effects for the patient, ADCs were heavily investigated in recent years and some were already approved by the FDA.[48] Antibodies consist of several units that are connected *via* disulfide bridges. The disulfides can be used to attach a linker, with or without a drug compound, to the antibody, without disturbing the antigen binding region. In Scheme 6 this is exemplified with a reduction of the disulfide **36** and subsequent thiol-yne reaction of the two formed thiols **37**.[49] The reaction product **38** provides a carboxylic acid that can be used to attach the payload and ultimately form the ADC.

Griebenow, Bräse *et al.* **2016**:

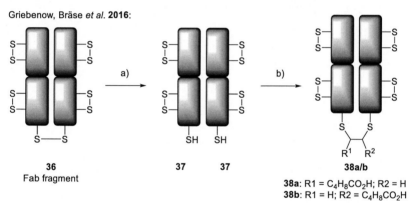

Scheme 6. Application of cystine bioconjugation in an antibody Fab fragment (**36**). To obtained product can be further modified at the carboxylic acid.[49] *a)* tris(2-carboxyethyl)phosphine hydrochloride, rt, 1 h; *b)* 6-heptynoic acid, lithium phenyl-2,4,6-trimethylbenzoylphosphinate (4 × 20 mol%), UV (365 nm), 0 °C, 4 × 1 h.

3.3 [1.1.1]Propellane and bicyclo[1.1.1]pentanes

The tricyclic structure of [1.1.1]propellane (**1**) has fascinated chemists long before the application of **1** in the synthesis of bioisosteres.[50] It was the first polyatomic compound with predictions of the vibrational spectrum and other properties prior to its synthesis.[51-53] Larger propellanes, e.g. the naturally occurring [3.3.3]propellane modhephene (**39**),[54] and propellanes based on other elements than carbon like **40**[55] will not be part of this thesis.

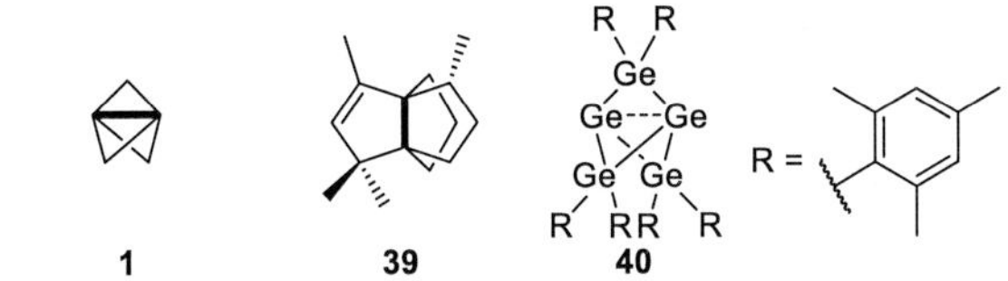

Figure 6. Structure of [1.1.1]propellane (**1**), natural product modhephene (**39**) and a germanium-based propellane **40**. The central bond is highlighted in **1** and **39**.

3.3.1 History and nomenclature of propellanes

The name propellane was coined by Ginsburg *et al.* in 1966 due to the propeller-like shape of the compounds.[56] To simplify the IUPAC nomenclature only the size of the rings is mentioned, without counting the central bond. This shortens the name tricyclo[1.1.1.0]pentane to [1.1.1]propellane. If the propellane contains rings of different size, the numbers are sorted in descending order, e.g. [4.3.1]propellane. Name suffixes are added in accordance with IUPAC, e.g. [4.4.4]propellatetraene. Polymers of **1** are called [*n*]staffanes with *n* being the number of BCP units.

With smaller ring sizes the ring strain of the molecule increases. The smallest propellane **1** is still surprisingly stable (bond energy ~60 kcal/mol for the central bond).[57-59] This observation led to several studies about the nature of the central bond in **1**. A suggested inverted σ-bond where the two smaller lobes of the sp hybrid molecular orbitals overlap has been the explanation for the stability for a while.[60] Wu *et al.* compared the CC bond in ethane to the central bond of **1** and found that they share a similar strength. However, the nature of the bond differs greatly. While in ethane a classically covalent bond is present, the two bridgehead carbons in **1** should be connected by a non-classical charge-shift bond, similar to the bond in molecular fluorine (F_2).[61]

Recently, Chaquin *et al.* used *in silico* models to calculate the binding energy of the inverted bond and concluded that the bonding only results from π-type (banana) interactions.[62] The true nature of this bond remains an interesting topic for further investigations.

The first synthesis of **1** was published by Wiberg and Walker in 1982 (Scheme 7).[57] Starting from the dicarboxylic acid **41** they synthesized 1,3-dibromobicyclo[1.1.1]pentane (**42**) in a Hunsdiecker reaction. The dibromide **42** was treated with *tert*-butyllithium in a pentane/Et$_2$O mixture. The details of the synthetic procedure and the yield are not reported, but the identity of **1** was proven by NMR spectroscopy, mass spectrometry and subsequent reactions.

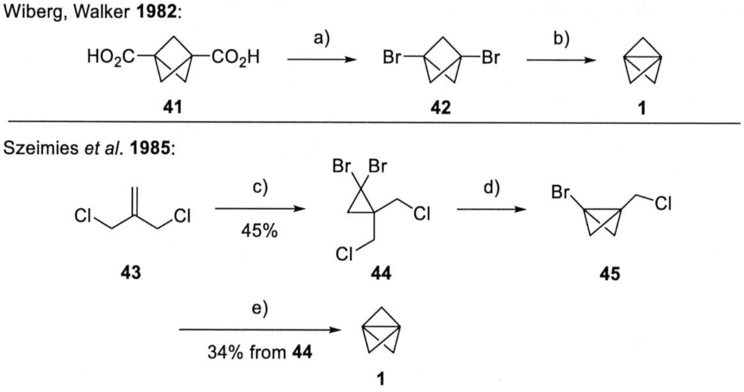

Wiberg, Walker **1982**:

Szeimies *et al.* **1985**:

Scheme 7. First synthesis of [1.1.1]propellane (**1**) by Wiberg and Walker[57] and the improved method by Szeimies *et al.*[63] *a)* **41**, Br$_2$, HgO; *b)* **42**, *t*-BuLi, pentane/Et$_2$O; *c)* **43**, CHBr$_3$, aq. NaOH, phase-transfer catalysis; *d)+e)* 1.00 equiv. **44**, 2.20 equiv. *n*-BuLi, pentane/Et$_2$O, –50 °C, 30 min.

Only three years later Szeimies *et al.* developed another strategy towards [1.1.1]propellanes through a dibromocyclopropane **44** (Scheme 7).[63] This product of a *gem*-dibromocarbene addition to **43** could be converted to **1** in a two-step reaction. The intermediate bicyclo[1.1.0]butane **45** allows the easier formation of **1** as the ring strain is formed in two smaller subsequent steps.[64] This method was further improved by several groups and is still the standard synthesis of [1.1.1]propellanes today.[65-66]

Cyclopropane **44** is commercially available nowadays and the lithiation can be performed with different organolithium reagents. Most common are methyllithium and phenyllithium. The volatile product **1** can be distilled and collected, together with the solvent, as a colorless solution. Solvent-free **1** can be prepared, but is prone to polymerization.[67] Before usage, the concentration of **1** in the solution needs to be determined, either by quantitative NMR[66] or by the quantitative reaction of an aliquot with thiophenol (see chapter 5.1).[68]

The above mentioned stability of **1** is of course limited to the present conditions. Usually, solutions of **1** are stored under inert atmosphere at low temperatures (–20 to –78 °C). At high temperatures (430 °C) the tricyclic compound rearranges to 1,2-dimethylenecyclopropane (**46**, Scheme 8).[69] If the BCP cation **47** is formed, e.g. by addition of acid, it rearranges to the methylenecyclobutyl cation **48** which can be attacked by a nucleophile (Scheme 8).[57] In the following chapter only reactions of **1** that lead to the formation of BCPs will be discussed.

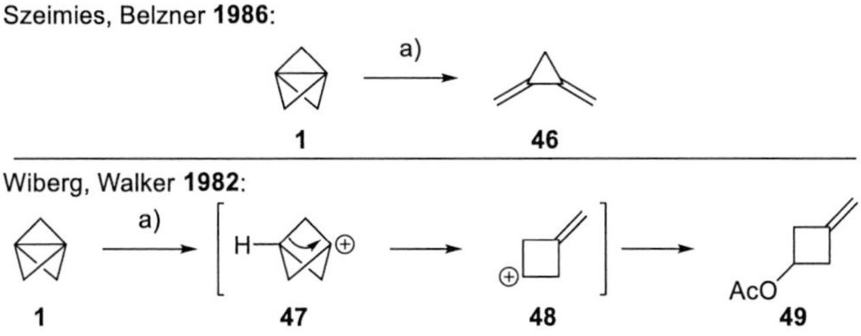

Szeimies, Belzner **1986**:

Wiberg, Walker **1982**:

Scheme 8. Rearrangements of **1** under thermal[69] and acidic conditions.[57] *a)* flow system, 430 °C; *b)* AcOH.

3.3.2   Synthesis of bicyclo[1.1.1]pentanes from [1.1.1]propellane

In this chapter a non-comprehensive summary of BCP syntheses from **1** is presented. For more details, several reviews should be considered.[35, 50, 70-71]

Reactions of free radicals with **1** were heavily investigated and cover the majority of published reactions of **1**. Even reactions of nucleophiles with **1** were considered to proceed through a radical mechanism.[70, 72] Among many other free radical reactions, Wiberg and Waddell described the addition of iodine to **1** (Scheme 9).[73] The reaction proceeds instantly and the product can be sublimed to obtain colorless crystals in 88% yield. The 1,3-diiodobicyclo[1.1.1]pentane (**50**) was shown to be a precursor for **1** similar to the dibromide **42** in the first synthesis of [1.1.1]propellane (see chapter 3.3.1).[67, 74-75] More interestingly, the diiodide **50** can be converted to the iodoazide **51** through a cationic intermediate.[75] The second iodide stabilizes the intermediate and after trapping with an azide anion product **51** can be obtained in 79% yield. The azide **51** can be applied in strain-promoted azide-alkyne cycloaddition reactions.[76]

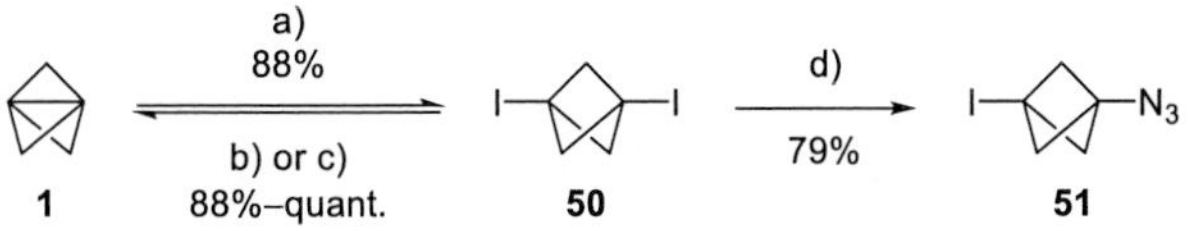

Scheme 9. Reaction of **1** with molecular iodine.[73] The diiodide **50** can be converted back to **1**[67, 74-75] or to azide **51**[75] which can be applied in click reactions.[76] *a)* 1.04 equiv. **1**, 1.00 equiv. I$_2$, rt; *b)* 1.00 equiv. **50**, 3.01 equiv. NaCN, DMSO, rt, 30 min, 88%; *c)* 1.00 equiv. **50**, 5.00 equiv. KOH, EtOH, rt, 2 h, quant.; *d)* 1.00 equiv. **50**, 2.56 equiv. NaOMe, 3.21 equiv. NaN$_3$, MeOH, rt, 4 d.

One of the first BCP building blocks that could easily be applied in the synthesis of bioisosteres was the dicarboxylic acid **54**. It can be obtained by photoinitiated radical addition of diacetyl (**52**) to **1** and subsequent hypobromite oxidation in multigram scale (Scheme 10).[77] The diacid **54** can be converted to the non-symmetrical BCP **55** through esterification and monohydrolysis.

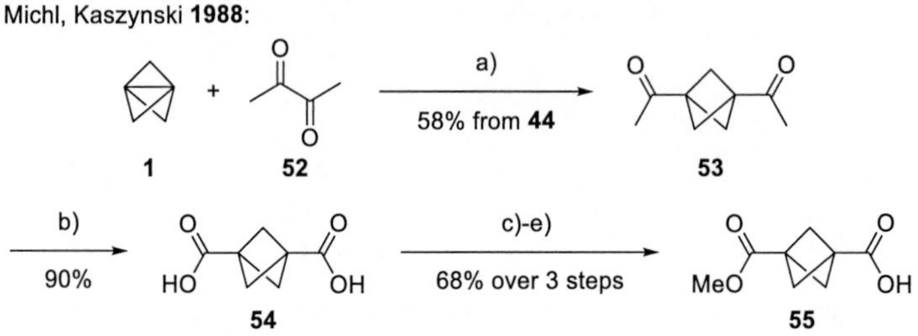

Scheme 10. Synthesis of dicarboxylic acid **54** and subsequent transformation to the non-symmetrical BCP building block **55**.[77] *a)* 1.00 equiv. **1**, 1.03 equiv. **52**, *hv* (UV, 450 W), 0 °C, 8 h; *b)* 1.00 equiv. **53**, 7.23 equiv. Br₂, 20.2 equiv. NaOH, 1,4-dioxane/H₂O, 0–50 °C, 7 h; *c)* 1.00 equiv. **54**, 3.95 equiv. SOCl₂, 74 °C, 10 h, 89% **54-(COCl)₂**; *d)* 1.00 equiv. **54-(COCl)₂**, 13.3 equiv. MeOH, 65 °C, 30 min, 99% **54-(CO₂Me)₂**; *e)* 1.00 equiv. **54-(CO₂Me)₂**, 1.00 equiv. NaOH, MeOH, 65 °C, 2.5 h, 78% **55**.

As commonly used organometallic reagents, Grignard compounds are widely applied and in many cases commercially available. The reaction of **1** with such compounds was investigated first by Szeimies *et al.*[72] and by de Meijere *et al.*[78] (Scheme 11). Both groups used similar conditions for the formation of BCP Grignard compounds **57/60** and applied the reagents in subsequent Kumada coupling reactions. Szeimies *et al.* focused on palladium as catalyst for the coupling while de Meijere *et al.* compared palladium to nickel. Shortly after these publications Szeimies *et al.* extended their methodology to the homocoupling of **57** to obtain [2]staffanes (not shown).[79]

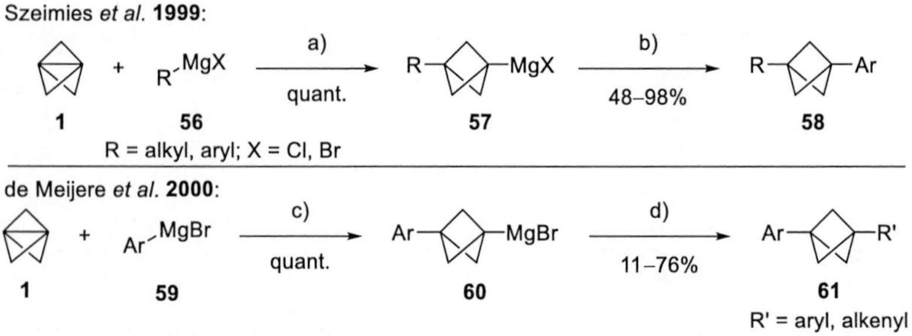

Scheme 11. Reactions of **1** with Grignard reagents and subsequent Kumada couplings.[72, 78] *a)* 1.00 equiv. **1**, 1.00 equiv. **56**, rt, 2–6 d; *b)* 1.08–1.66 equiv. **57**, 1.00 equiv. ArBr, 1.3–3.0 mol% PdCl₂(dppf), 1,4-dioxane, rt, 48 h; *c)* 1.13 equiv. **1**, 1.00 equiv. **59**, 35 °C, 3–7 d; *d)* 4.00 equiv. **60**, 1.00 equiv. R'I or R'Br, 9–11 mol% NiCl₂(dppe) or 2 mol% PdCl₂(dppf), Et₂O or THF, rt, 16–72 h.

Although de Meijere *et al.* extensively studied the reactions of **60**, including the reaction with electrophiles and transmetallation with zinc followed by Negishi couplings, the syntheses of aryl BCP Grignard reagents still had a major drawback. The reaction time was usually in the range of several days.[78] Knochel *et al.* could perform the conversion in a shorter reaction time of only a few hours by increasing the temperature to 100 °C (Scheme 12).[80] The elevated temperature required the use of pressure vials as the solvent was usually diethyl ether. The group applied the method in the synthesis of two bioisosteres for internal alkynes. However, no comparison of activities of the synthesized compounds and the original drugs has been published so far.

Knochel *et al.* **2017**:

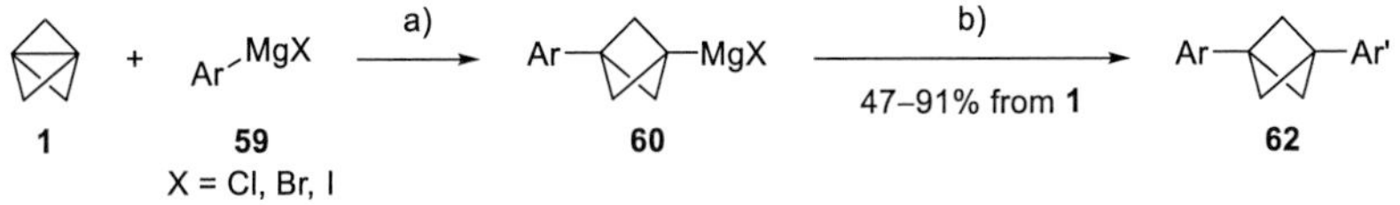

Scheme 12. Reaction of **1** with Grignard reagents **59** at higher temperatures in pressure vials. The intermediates **60** could be transmetallized with zinc and applied in Negishi couplings.[80] *a)* 1.00 equiv. **1**, 2.00 equiv. **59**, Et$_2$O, 100 °C, 45 min to 3 h; *b)* crude **60**, 2.20 equiv. ZnCl$_2$, THF, 0 °C, 5 min, then 2.08 equiv. Ar'X, 2 mol% PdCl$_2$(dppf)·CH$_2$Cl$_2$, 40–65 °C, 1–18 h.

As a second part of their study, de Meijere *et al.* also used alkyl iodides **63** to react with **1** in a radical fashion (Scheme 13).[78] The obtained alkyl BCP iodides **64** were lithiated with *tert*-butyllithium and either reacted with an electrophile (not shown) or transmetallated with zinc and applied in Negishi couplings. For the cross coupling reaction aryl triflates could be used instead of aryl halides with similar yields.

de Meijere *et al.* **2000**:

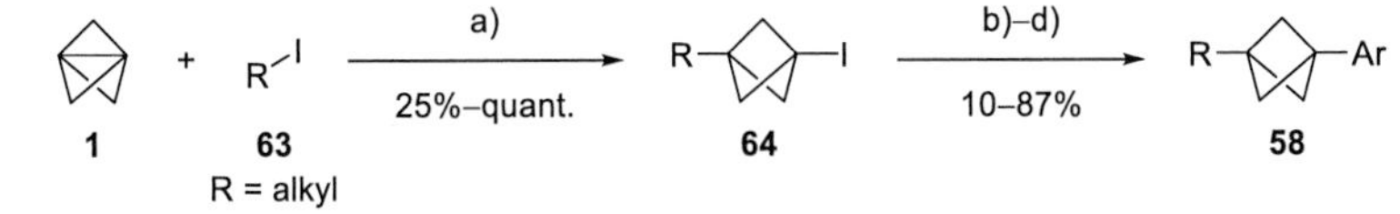

Scheme 13. Reaction of **1** with alkyl iodides **63** followed by lithiation, transmetallation and Negishi coupling.[78] *a)* 1.07 equiv. **1**, 1.00 equiv. **63**, 1.00 equiv. MeLi, Et$_2$O, –40 °C to rt, 24 h; *b)* 1.00 equiv. **64**, 2.00 *t*-BuLi, Et$_2$O, –78 °C, 1.5 h; *c)* crude **64-Li**, ZnCl$_2$, THF, –78 °C to rt, 1 h; *d)* 0.50–4.00 equiv. crude **64-ZnCl**, 1.00 equiv. ArX (X = Br, I, OTf), 2–10 mol% PdCl$_2$(dppf), Et$_2$O or THF, rt, 2–72 h.

The stoichiometric amount of methyllithium or the initiation with UV irradiation to generate alkyl radicals limited the scope of the alkyl BCP iodide synthesis. Anderson *et al.* therefore investigated triethylborane as radical initiator and performed atom transfer radical additions (ATRA) with activated alkyl iodides and bromides **65** (Scheme 14).[81] The milder conditions tolerated a variety of functional groups and the method could be applied to a nucleotide and a dipeptide. Through deiodination with tris(trimethylsilyl)silane (TTMSS) and catalytic triethylborane the terminal BCP compounds **67** could be obtained in good yields.

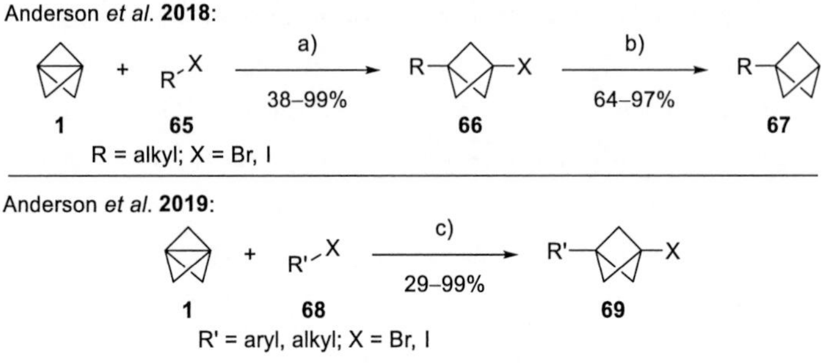

Anderson *et al.* **2018**:

R = alkyl; X = Br, I

Anderson *et al.* **2019**:

R' = aryl, alkyl; X = Br, I

Scheme 14. Atom transfer radical addition of halides **65/68** to **1**.[81-82] The alkyl BCP iodides **66** could be converted to the deiodinated terminal BCPs **67**. *a)* 1.10–2.00 equiv. **1**, 1.00 equiv. **65**, 1–10 mol% BEt$_3$, 0 °C or rt, 15 min to 15 h; *b)* 1.00 equiv. **66**, 1.30–2.00 equiv. TTMSS, 10 mol% BEt$_3$, MeOH, rt, 15 min to 2 h; *c)* 2.00 equiv. **1**, 1.00 equiv. **68**, 2.5 mol% *fac*-Ir(ppy)$_3$, *t*-BuCN, *hv* (blue LED), 25 °C, 0.5–24 h.

Recently, the same group expanded the reaction of halides with **1** by using photoredox catalysis to generate the radicals (Scheme 14).[82] They could successfully synthesize aryl, benzyl, primary and secondary alkyl BCP iodides **69** (and bromides in special cases) including complex natural product analogues. With appropriate substrates cyclizations could be performed followed by the reaction with **1** in a tandem reaction.

In another recent study, Walsh *et al.* reacted aryl dithianes **70** with **1** (Scheme 15).[83] The base induced reaction does presumably not proceed through a radical mechanism, as proposed for other nucleophiles before, but through an anionic opening of [1.1.1]propellane. With density functional theory (DFT) calculations the Walsh group compared the two-electron vs the one-electron pathway and found that the anionic intermediate is more likely to occur. The dithiane products **71** could be deprotected to the corresponding ketones **72** or converted to the difluoromethane derivatives (not shown).

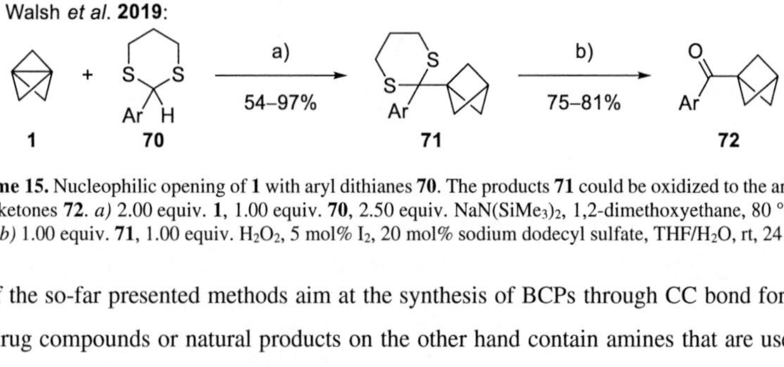

Walsh *et al.* **2019**:

Scheme 15. Nucleophilic opening of **1** with aryl dithianes **70**. The products **71** could be oxidized to the aryl BCP ketones **72**. *a)* 2.00 equiv. **1**, 1.00 equiv. **70**, 2.50 equiv. NaN(SiMe$_3$)$_2$, 1,2-dimethoxyethane, 80 °C, 16 h; *b)* 1.00 equiv. **71**, 1.00 equiv. H$_2$O$_2$, 5 mol% I$_2$, 20 mol% sodium dodecyl sulfate, THF/H$_2$O, rt, 24 h.

Most of the so-far presented methods aim at the synthesis of BCPs through CC bond formation. Many drug compounds or natural products on the other hand contain amines that are used for a modular synthesis. Therefore, many groups contributed to the synthesis of BCP amine (**76**) and its derivatives. The parent compound BCP amine (**76**) has been available for some time, but the application in an industrial process was limited due to the use of toxic and explosive reagents. Bunker *et al*. developed a three step procedure (from **1**) that could be scaled-up easily and therefore overcome this limitation in drug discovery (Scheme 16).[84] Through a manganese-catalyzed hydrohydrazination intermediate **74** could be obtained in quantitative yield. After deprotection and reduction the amine **76** could be synthesized in 80% yield.

Baran *et al*. could cut another step in the synthesis of **76** by using the 'turbo-amide' (Scheme 16).[66, 85] The group coined the reaction strain-release amination and applied it to different secondary amines including late-stage drug candidates.

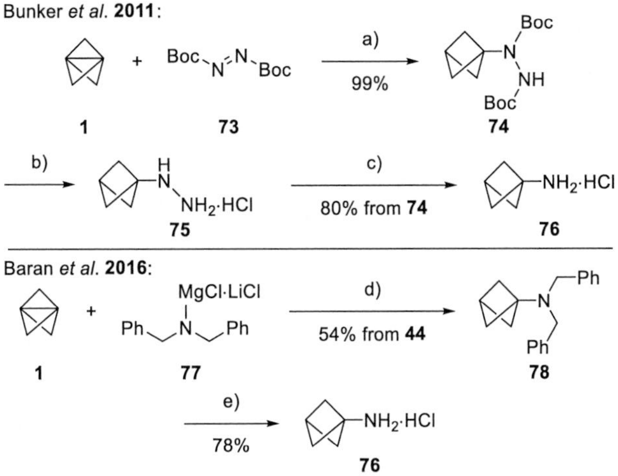

Scheme 16. Scalable syntheses of BCP amine (**76**) by Bunker *et al*.[84] and Baran *et al*.[66, 85] *a*) 1.00 equiv. **1**, 1.50 equiv. **73**, 1.00 equiv. PhSiH₃, 2 mol% Mn(dpm)₃, *i*PrOH/CH₂Cl₂, 0 °C, 21 h; *b*) 1.00 equiv. **74**, HCl, EtOAc, rt, 18 h; *c*) crude **75**, H₂ (3 bar), 10 mol% PtO₂, MeOH, 25 °C, 24 h; *d*) 1.00 equiv. **1**, 2.00 equiv. **77**, Bu₂O, 0–60 °C, 16 h; *e*) 1.00 equiv. **78**, H₂ (3.4 bar), 2.5 mol% Pd(OH)₂, MeOH, 30 °C, 12 h.

The progress in the synthesis of terminal BCP amines inspired other groups to aim for 3-substituted BCP amines (Scheme 17). Uchiyama *et al*. used a radical multicomponent carboamination to form 3-substituted BCP hydrazines **80** that could then be converted to the corresponding amines **82**.[86] The initial addition of a carbon-centered radical to **1** forms a BCP radical that can be trapped by di-*tert*-butyl azodicarboxylate (**73**). The group calculated by DFT that this trapping is kinetically favored over the oligomerization of the BCP radicals.

Gleason *et al.* expanded the method with 'turbo amides' and used the metallated BCP intermediate **84** for further modifications (Scheme 17).[87] By the addition of copper(I) iodide they could catalyze the reaction with alkyl halides (and triflates) to obtain 3-alkyl substituted BCP amines **85**.

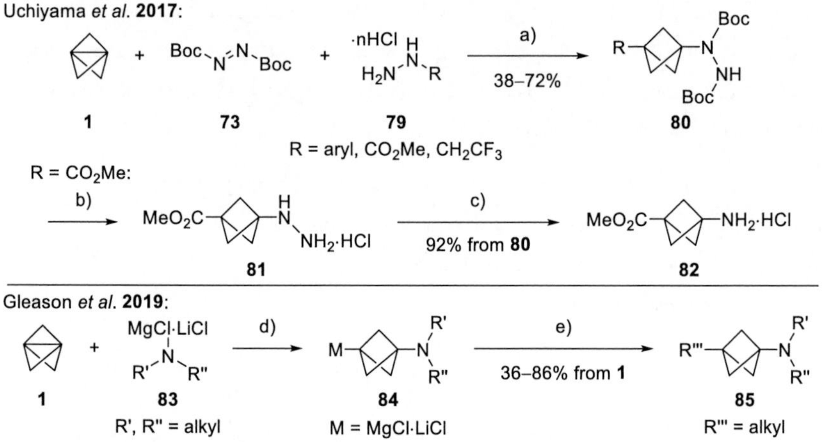

Uchiyama *et al.* **2017**:

R = aryl, CO$_2$Me, CH$_2$CF$_3$

R = CO$_2$Me:

Gleason *et al.* **2019**:

Scheme 17. Synthesis of 3-substituted BCP hydrazines **80** and amines **82/85** by Uchiyama *et al.*[86] and Gleason *et al.*[87] *a)* 1.00 equiv. **1**, 2.00 equiv. **73**, 2.00 equiv. **79**, 1.50 equiv. *tert*-butylhydroperoxide, n + 1.00 equiv. Cs$_2$CO$_3$, 5 mol% iron(II) phthalocyanine, CH$_3$CN, −20 °C, 1–3 h; *b)* 1.00 equiv. **80**, 10 equiv. HCl, EtOAc, rt, 36 h; *c)* crude **81**, H$_2$ (1 bar), 10 mol% PtO$_2$, MeOH, rt, 12 h; *d)* 1.00 equiv. **1**, 2.00 equiv. **83**, Et$_2$O/THF, rt, 16 h; *e)* crude **84**, 2.20 equiv. R'''X (X = Br, I, OTf), 10 mol% CuI, 50 °C, 24 h.

While the synthesis of BCP amines was intensively studied in recent years, the corresponding sulfur analogues have not been addressed until the beginning of this work. There are many ways to use sulfur-based radicals, e.g. in free-radical thiol-ene reactions, that could possibly be applied to [1.1.1]propellane as well. Besides the projects presented in this thesis, there was a recent publication by Riant *et al.* in which they investigated the radical addition of xanthates **86** to **1** (Scheme 18).[88]

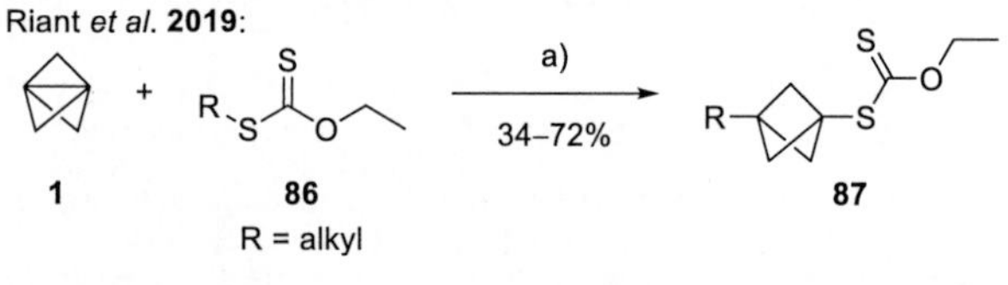

Riant *et al.* **2019**:

R = alkyl

Scheme 18. Synthesis of BCP xanthates **87** by radical exchange.[88] *a)* 2.00 equiv. **1**, 1.00 equiv. **86**, 10–20 mol% dilauroyl peroxide, dichloroethane, 80 °C, 8–10 h.

The obtained BCP xanthates **87** could possibly be further modified,[89] in this work the hydrolysis to the thiol was shown in one case (see SI of the publication).[88]

3.3.3   Synthesis of bicyclo[1.1.1]pentanes from other precursors

While it is less commonly applied, the synthesis of BCPs is still possible from other precursors than [1.1.1]propellane. The precursor for the first synthesis of **1** (see chapter 3.3.1) was synthesized by addition of dichlorocarbene (**89**) to bicyclo[1.1.0]butane (BCB) **88**, followed by oxidation to **91** and reduction of the two chlorides to **55** with TTMSS (Scheme 19).[90] A very similar strategy was used by Hirst *et al.* to synthesize a bioisostere of drug candidate Darapladib, an inhibitor of lipoprotein-associated phospholipase A2 (LpPLA₂).[91] They argued that a synthesis strategy towards **94** using the carbene was more suitable for large scale chemistry. The obtained bioisostere (not shown) showed improved physicochemical properties while maintaining the initial activity.

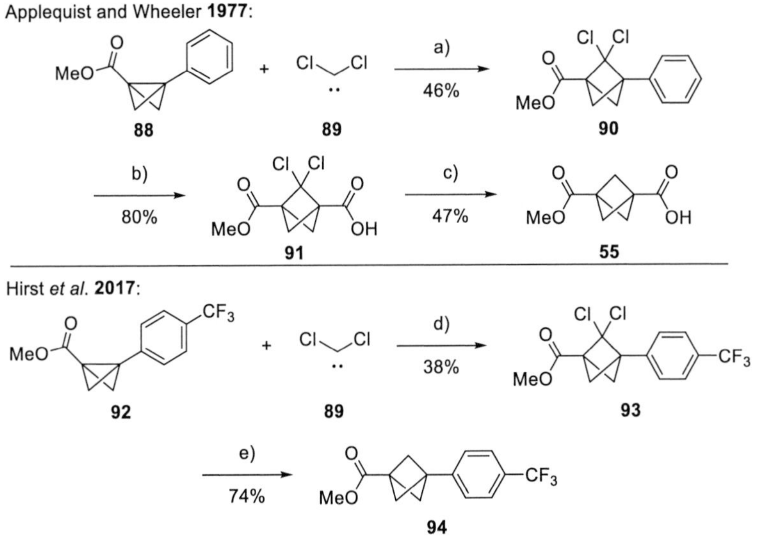

Scheme 19. Dichlorocarbene addition to bicyclo[1.1.0]butanes to obtain BCPs.[90-91] The carbene **89** in both cases was formed according to Fieser and Sachs.[92] *a)* 1.00 equiv. **88**, 2.00 equiv. **89** (generated from CCl₃CO₂Na), tetrachloroethylene/diglyme, reflux, 72 h; *b)* **90**, RuO₂, NaOCl, CCl₄/H₂O; *c)* **91**, Bu₃SnH; no further details of the procedures *a)–c)* available; *d)* 1.00 equiv. **92**, 5.00 equiv. **89** (generated from CCl₃CO₂Na), tetrachloroethylene/diglyme, 120–140 °C, 1 h; *e)* 1.00 equiv. **93**, 5.74 equiv. TTMSS, 1.19 equiv. 1,1'-azobis(cyclohexanecarbonitrile), toluene, 110 °C, 4 d.

A major challenge in the application of BCPs as bioisosteres is the substitution of BCPs at the 2-position. The dichlorocarbene addition already delivers such a compound, although both research groups reduced the chlorines for their purposes. Pharmacologically more interesting are 2-fluoro BCP derivatives, for reasons discussed in chapter 3.1.1. Two independent studies with 2,2-difluorinated BCPs were published only a few months apart (Scheme 20).[93-94] Both used difluorocarbene (**96**) addition to BCBs similar to the dichloro BCP synthesis above. The presented scope was similar in both publications, although Mykhailiuk *et al.* also applied the reaction to an alkenyl substituted BCB.[94] Monosubstituted substrates were unsuccessful in both studies.

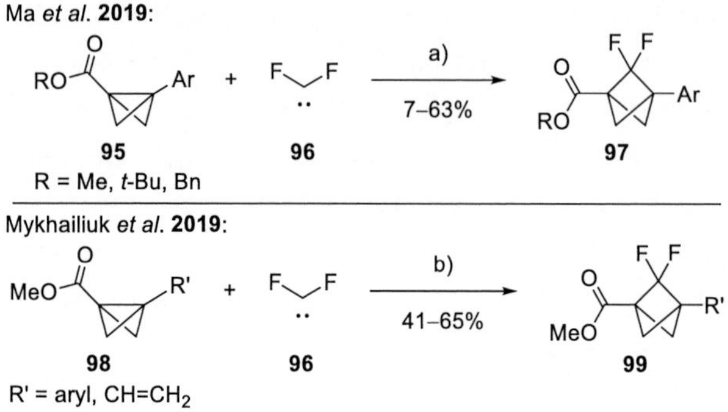

Ma *et al.* **2019:**

Mykhailiuk *et al.* **2019:**

Scheme 20. Difluorocarbene (**96**) addition to bicyclo[1.1.0]butanes to obtain BCPs.[93-94] *a*) 1.00 equiv. **95**, 3.00 equiv. **96** (generated from trimethylsilyl 2-fluorosulfonyl-2,2-difluoroacetate (TFDA)), 50 mol% NaF, 1,4-dioxane, 90 °C, 15 min; *b*) 1.00 equiv. **98**, 3.00 equiv. **96** (generated from CF₃Si(CH₃)₃)), 50 mol% NaI, THF, 65 °C, 4 h.

Another way to obtain 2-substituted BCP derivatives is the synthesis of 2-substituted [1.1.1]propellanes and application of the previously described methods (see chapter 3.3.2). There are only very few studies towards such structures and even less deal with functional groups at the 2-position, which would be desirable for applications as bioisosteres[35] or for bioconjugations. Szeimies *et al.* established a route to mono- and disubstituted [1.1.1]propellanes **103** through intramolecular carbene formation and subsequent cyclization (Scheme 21).[95] The substituents R and R' were introduced in **100** *via* the ketone RC=OR'. In a Corey-Kim type reaction a carbenium ion was formed and the bridge of the BCB was opened. The cation next to the bromide could be trapped by a chloride to give dihalide **101**. Lithiation and elimination of lithium chloride leads to the formation of carbene **102** that can add to the alkene and form the propellane **103**.

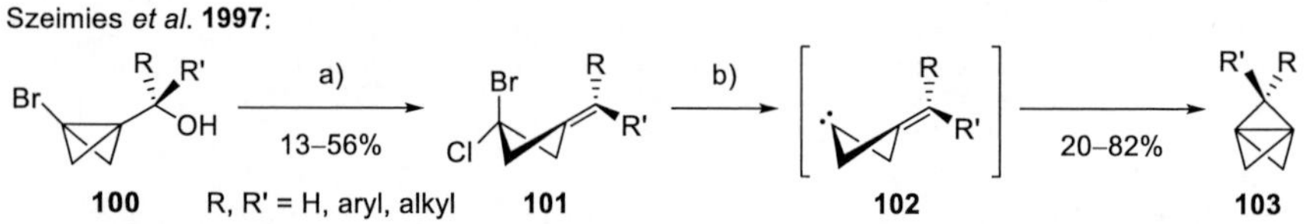

Szeimies *et al.* **1997:**

Scheme 21. Synthesis of 2-substituted [1.1.1]propellanes **103** from a methylenecyclobutane precursor **101** by intramolecular cyclopropanation. *a*) 1.00 equiv. **100**, 1.50 equiv. *N*-chlorosuccinimide, 1.50 equiv. Me₂S, CH₂Cl₂, −25 °C to 25 °C, 3–5 h; *b*) 1.00 equiv. **101**, 1.30 equiv. MeLi, Et₂O, −25 °C to rt, 3–18 h.

4 Objective

Bicyclo[1.1.1]pentanes (BCPs) recently gained enormous interest in medicinal chemistry as bioisosteres for *para*-substituted benzene, alkyne or *tert*-butyl moieties. Their rigid scaffold and three-dimensional structure can lead to improved physicochemical properties of drug candidates, while maintaining the biological activity of the compound. Besides numerous contributions, the synthesis of BCPs is still challenging in many cases and novel methods are of great interest to employ this motif in structure-activity relationship (SAR) studies.

The standard precursor in the synthesis of BCPs is [1.1.1]propellane (**1**), which reacts with free radicals or nucleophiles in most cases. The availability of methods for CC and CN bond formations with **1** expanded during the last two decades and 1-substituted as well as 1,3-disubstituted BCPs of this kind are generally available, with a few exceptions. The formation of CS bonds with **1** on the other hand was somehow overlooked in the field, although sulfur offers ideal properties for radical or nucleophile reactions. Therefore, the aim of this thesis was to investigate and develop methods for the reaction of sulfur-based groups (thiols and disulfides) with **1** (Scheme 22).

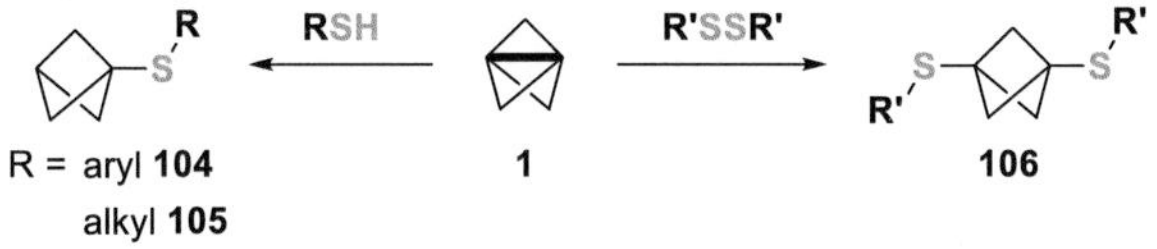

Scheme 22. Planned reactions of [1.1.1]propellane (**1**) with thiols and disulfides to obtain BCP sulfides **104–106**.

For a successful application the properties of the products **104–106**, e.g. polarity, should be tunable. The oxidation and imination of the BCP sulfides should be part of this thesis to obtain a variety of functional groups next to the BCP (Scheme 23). The synthesis of sulfoximines **108** should be of special interest, as these groups themselves are often employed as bioisosteres and show biological activities in many cases.

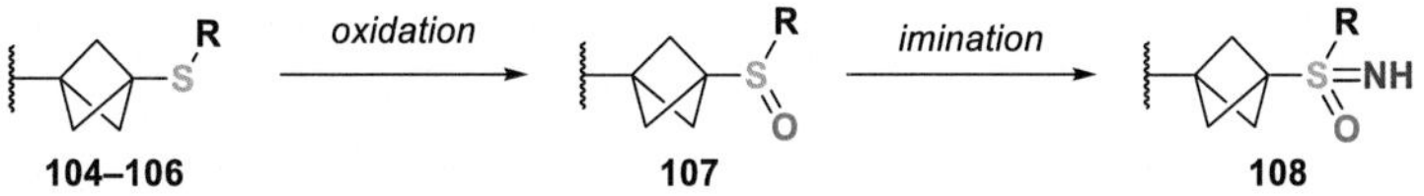

Scheme 23. Examples for possible modifications of the BCP sulfides **104–105** through oxidation to sulfoxides **107** and imination to sulfoximines **108**.

As the synthesis and handling of the volatile **1** requires Schlenk techniques and large amounts of organolithium reagents, the application in drug discovery might be limited. A bench-stable precursor would offer a solution to this and enable broad utilization of sulfur BCPs. This work aimed at the synthesis of such a precursor and the application in a variety of reactions.

5   Results and discussion

With rising interest in BCPs as bioisostere motifs in medicinal chemistry, the demand for novel synthetic methods increased as well. The synthesis of sulfur BCPs, e.g. BCP sulfides, has not been addressed so far. This thesis summarizes the results of four different projects around this area. The thiol addition to [1.1.1]propellane will be discussed first, as it enables access to a variety of terminal BCP sulfides in a fast and simple manner. As a next step the insertion of [1.1.1]propellane into disulfide bonds will be presented. In the third chapter different modifications of the obtained BCP sulfides will be shown. The last part deals with a novel bench-stable precursor for several sulfur BCP structures.

5.1   Thiol addition to [1.1.1]propellane[1]

When Szeimies *et al.* published their new route to [1.1.1]propellanes like **1** through the dibromocyclopropane precursor **44** in 1985 (see chapter 3.3.1), they mentioned the addition of a thiol to a propellane for the first time. The reaction of thiophenol (**110a**) with the bridged propellane **109** afforded the sulfide **111** (Scheme 24).[63] A yield was not indicated in this report. Wiberg and Waddell caught up on this and included the reaction of **110a** with **1** in their extensive paper 'Reactions of [1.1.1]Propellane'.[73] Their reported yield of 98% led to the use of thiophenol as a tool compound to determine concentrations of propellane solutions.[68] Besides this application the reaction remained neglected for the next two decades.

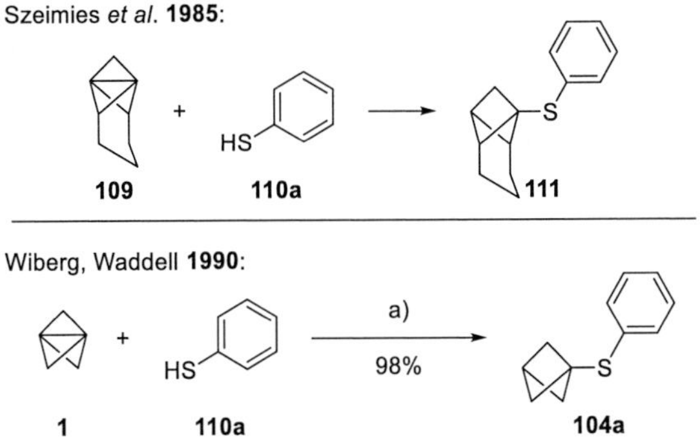

Scheme 24. First reports of thiol additions to [1.1.1]propellanes by Szeimies *et al.*[63] and Wiberg and Waddell.[73] *a)* 1.04 equiv. **1**, 1.00 equiv. **110a**, *n*-pentane/Et$_2$O, rt, 10 min.

---

[1] Excerpts of this chapter were already published in
R. M. Bär, S. Kirschner, M. Nieger, S. Bräse, *Chem. Eur. J.* **2018**, *24*, 1373–1382.

In this chapter the reaction of **1** with different thiols and one selenol will be discussed. The thiol scope was investigated and the discussion of the results is divided into aromatic and aliphatic thiols. Additionally, an isotope labelling experiment and competitive reactions were performed to gain insight into the reaction mechanism.

5.1.1 Aromatic thiols

As the addition of thiophenol (**110a**) to **1** worked in a quantitative fashion (Scheme 24, bottom), it was an obvious step to try other aromatic thiols in this reaction (Table 2). The reaction showed very good functional group tolerance and a variety of substituted aromatic thiols could be applied. The effects of different substituents will be discussed further below.

In a typical procedure the thiol was dissolved in diethyl ether before adding the propellane solution. Only in the case of 4-nitrothiophenol (**110g**) tetrahydrofuran (THF) was used instead (entry 6), as the thiol did not dissolve in diethyl ether. The thiols with carboxylic acid groups (**110k** and **110o**, entries 10 and 14) dissolved very poorly in diethyl ether. In an attempt without additional solvent, besides the solvent of the propellane solution, the products could be formed successfully. For those reactions the reaction time was increased from 15 min to 1 h and the disappearance of the solid thiol in the reaction mixture was an indication for the finished reaction.

For almost all products the purification could simply be done in one step. Residual thiol was removed by washing with a 1 M NaOH solution and after evaporation of the organic solvent pure product was obtained. Only products with acidic protons like sulfides **104k** and **104o** were purified *via* column chromatography. For these two BCP sulfides single crystals were obtained and the structures could be confirmed by X-ray diffraction (Figure 7).

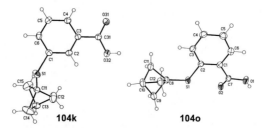

Figure 7. Structures of **104k** and **104o** determined by single-crystal X-ray diffraction. Displacement parameters are drawn at 50% probability level. Crystallographic data can be found in chapter 7.4.

Table 2. Addition of aromatic thiols **110** to [1.1.1]propellane (**1**). *a)* 1.00 equiv. **1**, 1.00–1.25 equiv. **110**, *solvent*, rt, 15 min. [a] Isolated yield; [b] part of master thesis;[96] [c] reaction time 1 h.

Entry	Thiol	R	Solvent	Product	Yield [%][a]
1	110b	4-Cl	Et$_2$O	104b	87[b]
2	110c	4-Br	Et$_2$O	104c	65
3	110d	4-Me	Et$_2$O	104d	87
4	110e	4-*t*-Bu	Et$_2$O	104e	79
5	110f	4-OMe	Et$_2$O	104f	90
6	110g	4-NO$_2$	THF	104g	47
7	110h	4-NH$_2$	Et$_2$O	104h	12[b]/54
8	110i	3-Cl	Et$_2$O	104i	72
9	110j	3-NH$_2$	Et$_2$O	104j	64
10	110k	3-CO$_2$H	–	104k	53[c]
11	110l	2-Cl	Et$_2$O	104l	58
12	110m	2-Me	Et$_2$O	104m	57
13	110n	2-OH	Et$_2$O	104n	61[b]
14	110o	2-CO$_2$H	–	104o	30[c]
15	110p	2,6-Cl$_2$	Et$_2$O/–	104p	50[b]/65
16	110q	3,5-Cl$_2$	Et$_2$O	104q	60
17	110r	2,4,6-Me$_3$	Et$_2$O	104r	84

The sulfides **104b**, **104n** and **104p** were already synthesized in the course of my master thesis.[96] In a repetition of **104p** without the addition of any solvent the yield could be increased from 50% to 65% (entry 15). It seems as the higher concentration of the thiol **104p** was beneficial for the conversion to the desired product. The synthesis of **104h** was repeated after my master thesis as well.[96] With a fresh batch of 4-aminothiophenol (**110h**) the product could be obtained in 54% instead of 12% (entry 7).

In some cases, the method for the preparation of [1.1.1]propellane (**1**) played a crucial role for the thiol addition (see chapter 3.3.1). If methyllithium was used for the synthesis of **1**,[68] bromomethane was formed as a by-product. This volatile compound (boiling point 4 °C)[97] was distilled together with **1** and the solvent. Activated thiols like **110f** reacted with methylbromide at the given reaction conditions and a product mixture was obtained. The mixture could be separated *via* column chromatography, but a larger excess of the thiol was necessary to achieve satisfactory yields of the desired BCP sulfides.

A control experiment was performed to confirm this reactivity (Scheme 25). Thiophenol (**110a**) was added to a propellane solution containing bromomethane (**112**) and the reaction mixture was stirred for 15 min at room temperature. Then, 4-methoxythiophenol (**110f**) was added and the mixture was stirred for another 15 min. After the usual work-up the two products **104a** and **113** were obtained as a mixture (see ^{1}H NMR spectra in chapter 7.3)

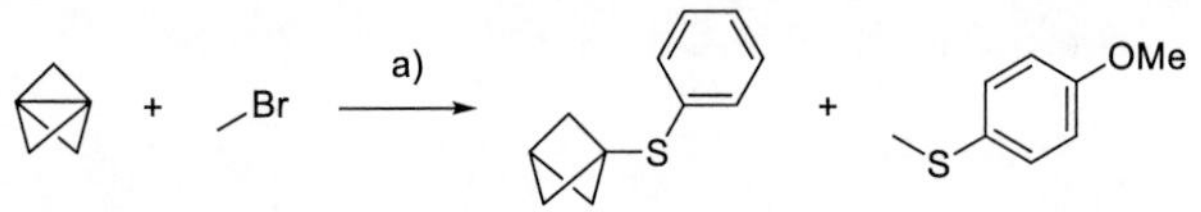

Scheme 25. Control experiment to confirm the difference in reactivity of aromatic thiols towards bromomethane (**112**). The yield was not determined. The ^{1}H NMR spectrum of the resulting mixture can be found in chapter 7.3 Selected NMR spectra. *a)* 1.00 equiv. **1**, 1.00 equiv. **110a**, Et$_2$O, rt, 15 min, then 1.00 equiv. **110f**, rt, 15 min.

To overcome the issue of methylation, the synthesis of **1** was performed with phenyllithium[66] instead of methyllithium. The by-product bromobenzene remained in the residue during the distillation due to the higher boiling point (156 °C)[98] and a pure [1.1.1]propellane solution was obtained.

With this [1.1.1]propellane (**1**) solution a variety of aromatic thiols **110** could be applied in the reaction (see Table 2) and some conclusions could be drawn from the obtained yields according to the substituent. Inductive effects seemed to have a minor influence on the yield of the corresponding sulfides **104**. Electron-withdrawing groups (EWG) like halides and electron-donating groups (EDG) like alkyl groups gave similar yields. Only the bromide **104c** and the dichlorides **104p** and **104q** showed a slight decrease in yield. In contrast, the mesomeric effects had a stronger influence on the reaction. A positive mesomeric effect (+M) for the methoxy-substituted **104f** led to an excellent yield of 90%. Negative mesomeric effects (–M) led to a decrease in yield to 53% and 30% for the carboxylic acids **104k** and **104o**, respectively. The amino-substituted products were obtained in lower yields. These compounds were found to be unstable at ambient conditions. Products **104h** and **104j** decomposed after a few days at room temperature without any inert atmosphere. The position of the substituent played an important role as well. *Ortho*-substituted thiols always led to lower yields compared to their *meta*- or *para*-homologues, which can be explained with steric hindrance.

An extension to heteroaromatic thiols was not successful (Scheme 26). Both reactions of **1** with 2-mercaptopyridyl **114** and with tetrazole thiol **116** led to complex product mixtures. In the case of tetrazole thiol **116** the rearranged product **117** could be isolated and identified by NMR (see spectra in chapter 7.3).

The desired tetrazole product **118** is in theory, after oxidation to sulfone **119**, an interesting building block for radical sulfone cross-couplings. Baran *et al.* developed the methodology for this reaction and showed its application for primary, secondary and fluorinated tertiary aliphatic sulfones (Scheme 27).[99] There's no comment in the publication about other tertiary aliphatic sulfones, but BCP would presumably be an interesting substrate as the intermediate radical **120** should be relatively stable.

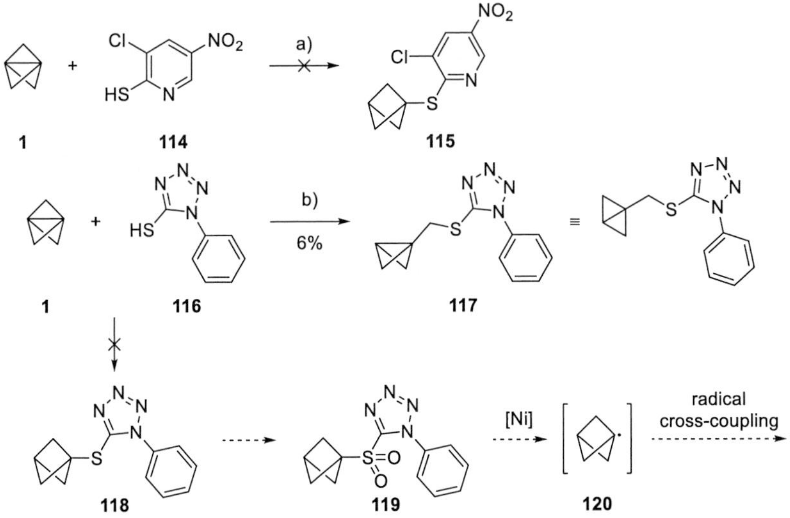

Scheme 26. Unsuccessful thiol additions to **1**. In the reaction with 2-mercaptopyridyl **114** a complex product mixture was obtained and no desired product could be found. The reaction with tetrazole thiol **116** led, among other non-identified products, to BCB **117**. The NMR spectra used for the structure elucidation of **117** can be found in chapter 7.3 Selected NMR spectra. *a)* 1.00 equiv. **1**, 1.20 equiv. **114**, THF, rt, 15 min; *b)* 1.00 equiv. **1**, 1.28 equiv. **116**, THF, rt, 15 min.

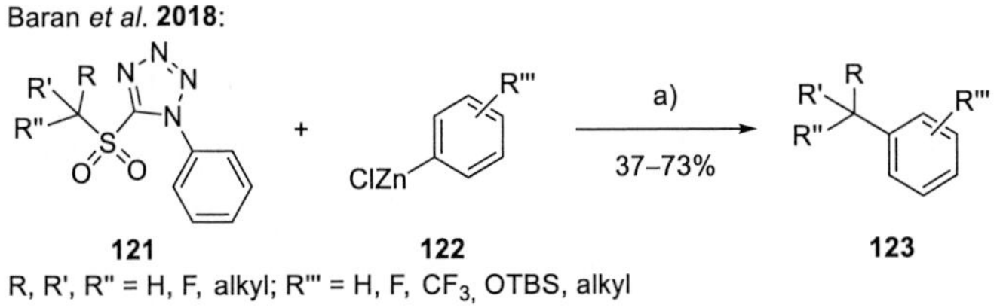

Baran *et al.* **2018**:

R, R', R'' = H, F, alkyl; R''' = H, F, CF$_3$, OTBS, alkyl

Scheme 27. Cross-coupling reaction of sulfones **121** and aryl zinc reagents **122**.[99]

After the success with most of the applied thiols, the reaction should be extended to selenols as well. There is precedence for reactions of [1.1.1]propellane (**1**) with selenocompounds (Scheme 28). Pellicciari *et al.* used a radical addition of acetyl phenylselenide (**124**) to **1**,[100] similar to the addition of **124** to alkenes.[101] They could also extend the reaction to phosphonoselenoate **126** and obtain 1-phenylseleno-3-diethoxyphosphonobicyclo[1.1.1]pentane (**127**).

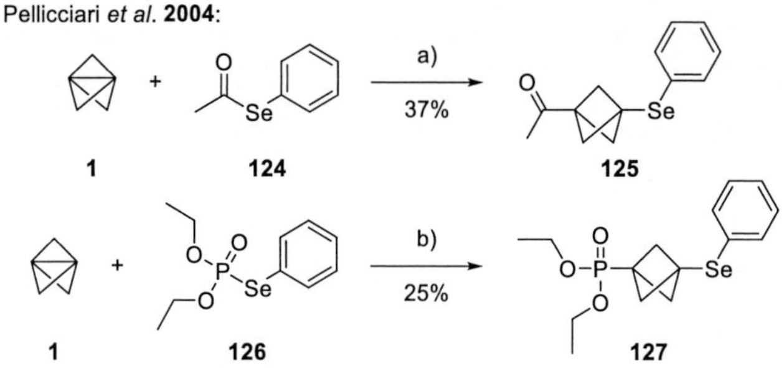

Pellicciari *et al.* **2004**:

Scheme 28. Examples for reactions of **1** with selenocompounds.[100] *a)* 1.00 equiv. **1**, 2.00 equiv. **124**, 5 mol% (*t*-BuO)₂, *hv*, 0 °C, 24 h; *b)* 1.00 equiv. **1**, 3.00 equiv. **126**, 5 mol% (*t*-BuO)₂, *hv*, 0 °C, 24 h.

The terminal BCP selenide **130** had previously been synthesized by Della *et al.* through a Barton decarboxylation of BCP carboxylic acid (**128**) as shown in Scheme 29. However, this starting material is only accessible through a multi-step procedure[102] or through a challenging lithiation of sulfide **104a**[73] and subsequent reaction with carbon dioxide (both not shown). The addition of selenols to **1** would enable facile access to BCP selenides.

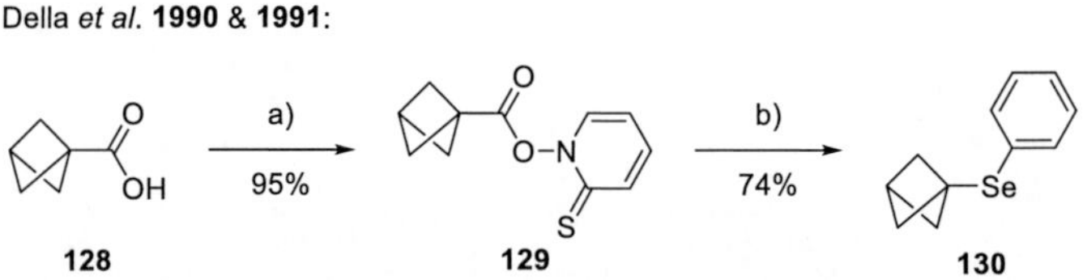

Della *et al.* **1990 & 1991**:

Scheme 29. Previous synthesis of BCP selenide **130** from carboxylic acid **128**. *a)* 1.00 equiv. **128**, 1.04 equiv. 2-mercaptopyridine *N*-oxide, 1.00 equiv. dicyclohexylcarbodiimide, CH₂Cl₂, 5 °C, 1.5 h, *dark*;[103] *b)* 1.00 equiv. **129**, 1.30 equiv. diphenyldiselenide, benzene, *hv*, rt, 25 min.[104]

Applying the reaction conditions of the thiol addition, benzeneselenol (**131**) reacted with **1** in a quantitative fashion (Scheme 30). A competitive reaction with **110a** and **131** revealed that the selenol reacted faster with propellane than the thiol (see ¹H NMR spectra in chapter 7.3). This is in accordance with the literature, as selenyl radicals are formed more easily than the corresponding thiyl radicals.[105]

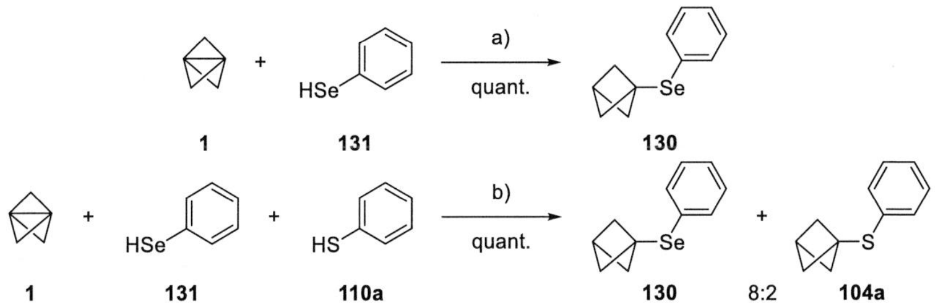

Scheme 30. Addition of benzeneselenol (**131**) to **1** and competitive reaction of selenol **131** and thiol **110a** with **1**. The yield was determined by ^{1}H NMR, the spectrum of the resulting mixture can be found in chapter 7.3 Selected NMR spectra. *a)* 1.00 equiv. **1**, 1.01 equiv. **131**, Et$_2$O, rt, 15 min; *b)* 1.00 equiv. **1**, 1.00 equiv. **131**, 1.00 equiv. **110a**, Et$_2$O, rt, 15 min.

5.1.2 Aliphatic thiols

Although the reaction of thiophenol (**110a**) with [1.1.1]propellane (**1**) was known for more than 30 years,[63] no alkyl thiol addition to **1** had been reported until the beginning of this project. With the same reaction conditions as for the aromatic thiol addition to **1** a variety of alkyl BCP sulfides **105** were successfully synthesized (Table 3).

The products with small alkyl substituents **105a–d** were obtained in poor to very good yields (entries 1–4). These four products are volatile and therefore a partial loss of the desired compounds during the purification is probable and the obtained yields varied in repeated reactions. Due to the volatile and non-polar nature of these products high-resolution mass spectrometry was not possible with the available methods. To confirm the proposed product structures GC-MS and NMR spectroscopy was used. The thiol **132e** was used as a mixture of isomers in the reaction and therefore the product **105e** was obtained as a mixture of isomers as well (entry 5). The ^{1}H NMR spectrum of this mixture is shown in chapter 7.3 Selected NMR spectra. The products **105f**, **105g** and **105j** were already part of my master thesis.[96] The synthesis of **105f** was repeated and the yield could be reproduced (entry 6). The functional group tolerance of the alkyl thiol addition was investigated and hydroxyl groups (**105g**), carboxylic acids (**105i**), protected amines (**105h** and **105j**) and esters (**105k**) proved to be suitable for this reaction (entries 7–11). Also a tri-*iso*-propylsilyl (TIPS) derivative **105l** could be synthesized in a quantitative fashion (entry 12).

Table 3. Addition of aliphatic thiols **132** to [1.1.1]propellane (**1**). *a)* 1.00 equiv. **1**, 1.00–1.25 equiv. **132**, *solvent*, rt, 15 min. [a] Isolated yield; [b] obtained as a mixture of isomers; [c] part of master thesis.[96]

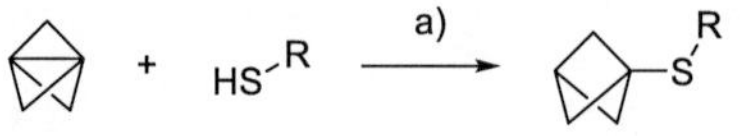

Entry	Thiol	R	Solvent	Product	Yield [%][a]
1	132a	*n*-Pr	Et$_2$O	105a	81
2	132b	*i*-Pr	Et$_2$O	105b	35
3	132c	*n*-Bu	Et$_2$O	105c	67
4	132d	*t*-Bu	Et$_2$O	105d	68
5	132e	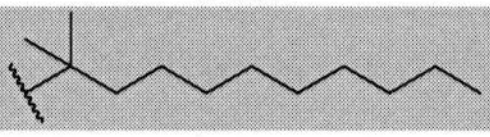 ⋀⋀⋀⋀⋀	Et$_2$O	105e	99[b]
6	132f	Bn	Et$_2$O	105f	52[c]/53
7	132g	(CH$_2$)$_2$OH	Et$_2$O	105g	77[c]
8	132h	(CH$_2$)$_2$NHFmoc	THF	105h	49
9	132i	CH$_2$CO$_2$H	Et$_2$O	105i	66
10	132j	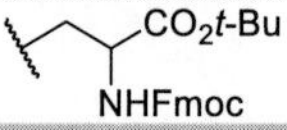 CO$_2$*t*-Bu / NHFmoc	THF	105j	45[c]
11	132k	(CH$_2$)$_2$CO$_2$Me	Et$_2$O	105k	99
12	132l	Si(*i*-Pr)$_3$	Et$_2$O	105l	99

With **105h** and **105j** novel building blocks for peptides and peptoids were provided for the Bräse group. They are expected to be suitable for solid phase synthesis following the Fmoc strategy.[106] Using the unprotected cysteamine (**132m**) or reacting **1** directly with peptoids like **132n** or **132o** was not successful and no desired product could be observed (Figure 8).

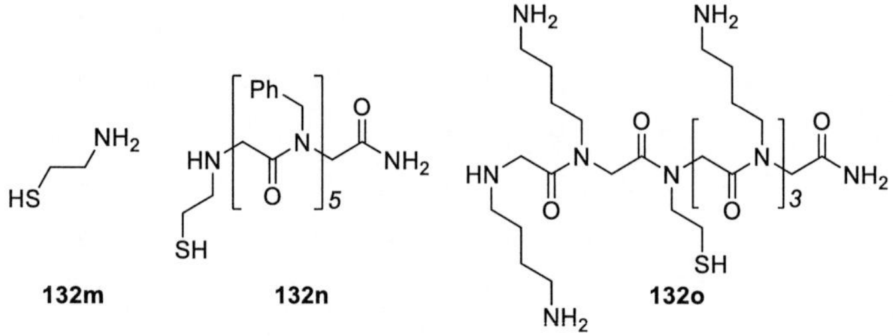

Figure 8. Unsuccessful substrates for the alkyl thiol addition to **1**. The peptoids **132n** and **132o** were provided by Anne Schneider (Bräse group).

During my master thesis dithiols were already used to synthesize bis(bicyclo[1.1.1]pentylsulfides) (not shown).[96] Another kind of bis-BCP sulfide should be obtained through the addition of hydrogen sulfide (**133**) to **1** (Scheme 31). The standard thiol addition procedure did not lead to any product formation. However, by irradiation with UV light (256 nm) the reaction proceeded smoothly in 15 min and the volatile product **134** could be obtained in quantitative yield.

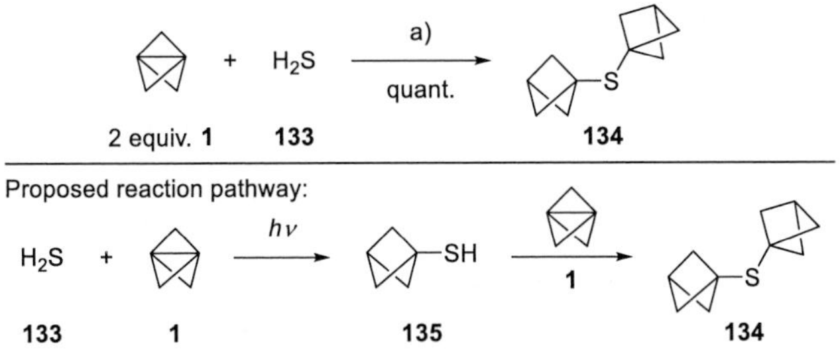

Scheme 31. Reaction of hydrogen sulfide (**133**) with **1**. H_2S only reacted with **1** after irradiation with UV-light. Presumably, BCP thiol (**135**) is formed in a first step and reacts with another equivalent of **1**. *a)* 2.00 equiv. **1**, 1.00 equiv. **133** in THF, *hv* (256 nm, 500 W), rt, 15 min.

The bis-BCP sulfide **134** suggests a two-step reaction with BCP thiol (**135**) as intermediate. Attempts to obtain this intermediate by stoichiometric control with a 1:1 ratio of H_2S and **1** were not successful. Other products of the thiol addition, like **105f** and **105l** should be suitable precursors of **135**. However, no deprotection of BCP thiol through hydrogenation of **105f** or fluoride-based desilylation of **105l** was successful (Scheme 32). Thiol **135** would be a useful intermediate in the synthesis of a variety of sulfur-based BCPs and apparently it is commercially available. Enamine Ltd. offers the thiol for ~2 000 $/g, but does not provide any information about physical state, pH, boiling point or relative density.[107]

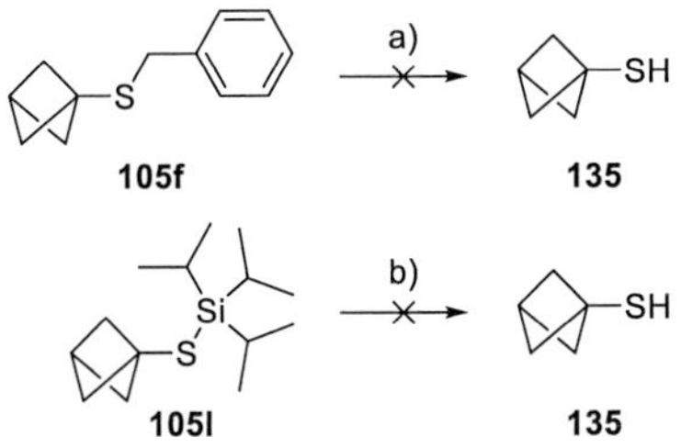

Scheme 32. Unsuccessful attempts to synthesize BCP thiol (**135**) from BCP sulfides **105f** and **105l**. *a)* 1.00 equiv. **105f**, H_2 (1 bar), Pd/C, EtOAc, rt, 16 h. *b)* 1.00 equiv. **105l**, 1.00 equiv. tetra-*n*-butylammonium fluoride, THF, 0 °C to rt, 30 min.

5.1.3 Mechanistic investigations

Thiols can react as radicals or nucleophiles, depending on the reaction conditions. Szeimies *et al.* proposed a radical chain process for the thiol addition to **1**.[63] This seems plausible, as the fast reaction of thiophenoxy radicals with **1** is well-known.[108] As a first step to gain more insight into the reaction pathway deuterated thiophenol (**110a**$_d$) was used in the reaction (Scheme 33). The product **104a**$_d$ was obtained with full incorporation of deuterium at the C3-position. This deuterium could either be from the very same thiophenol that attacked the propellane or from a second equivalent of thiophenol. To clarify this, a mixture study was performed with **110a**$_d$ and **110f**. A product mixture with four BCP-sulfides was obtained, containing deuterated and non-deuterated products from both thiols (see ^{1}H NMR spectra in chapter 7.3). This result supports the radical chain process as **104f**$_d$ can only be formed from a radical (or anionic) intermediate **136**. In a reaction with (2,2,6,6-tetramethylpiperidin-1-yl)oxyl (TEMPO, **137**) as a radical scavenger only traces of the desired product **104a** could be detected (Scheme 34). In a control experiment thiophenol (**110a**) and TEMPO (**137**) were mixed and stirred with the same reaction conditions. The formation of disulfide **138a** and adduct **139** were detected by GC-MS and NMR (see ^{1}H NMR spectra in chapter 7.3).

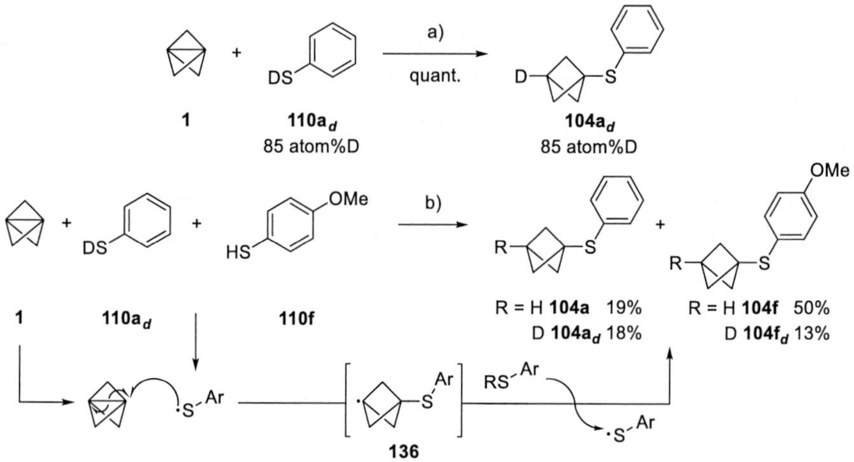

Scheme 33. Isotope labeling experiments performed to investigate the mechanism of the thiol addition to **1**. The yield of the mixture study was determined by ^{1}H NMR, the spectrum of the resulting mixture can be found in chapter 7.3 Selected NMR spectra. *a)* 1.09 equiv. **1**, 1.00 equiv. **110a**$_d$, rt, 15 min; *b)* 1.00 equiv. **1**, 1.00 equiv. **110a**$_d$, 1.00 equiv. **110f**, Et$_2$O, rt, 15 min.

The radical nature of this reaction was assumed based on literature and is in accordance with the obtained results. A nucleophilic opening of propellane was recently proposed by Walsh *et al.* for the reaction with 2-aryl-1,3-dithianes.[83] However, as the thiol addition does not require the addition of base and with the suppression of the reaction by adding TEMPO the radical chain reaction seems most likely.

Interestingly, the 4-methoxythiophenoxy radical (from **110f**) seems to be formed faster and/or to react faster with **1** as the corresponding yield of the sulfide is higher. This observation will be picked up in chapter 5.2.4 for discussion.

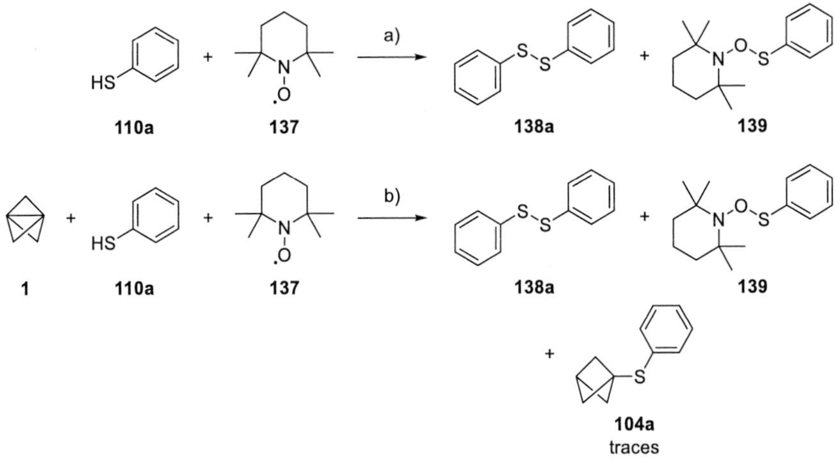

Scheme 34. Reactions with TEMPO (**137**) as radical scavenger. A control experiment with **110a** and **137** led to the formation of disulfide **138a** and **139**. In the thiol addition with TEMPO only traces (< 10%) of the product **104a** could be detected. Yields were not determined, products were identified by GC-MS. The ^{1}H NMR spectrum of the resulting mixture can be found in chapter 7.3 Selected NMR spectra. *a)* 1.00 equiv. **110a**, 1.00 equiv. **137**, Et$_2$O, rt, 16 h; *b)* 1.00 equiv. **1**, 1.00 equiv. **110a**, 1.00 equiv. **137**, Et$_2$O, rt, 16 h.

Overall, the thiol addition to **1** is a versatile tool for the synthesis of terminal BCP sulfides. Aryl and alkyl thiols with a variety of functional groups were used in the simple reaction and the products were obtained purely after one washing step in almost every case. This reaction enables the use of BCP sulfides as bioisosteres in drug discovery. With a suitable 2-substituted [1.1.1]propellane bearing a functional group, a bioconjugation with cysteines could be possible as well.

5.2   Insertion of [1.1.1]propellane into disulfide bonds[2]

The majority of bioisosteric replacements with BCP need a 1,3-disubstituted BCP to mimic a *para*-substituted benzene or alkyne. With the experiences from the thiol addition to **1** an extension to 1,3-bissulfide BCPs was the aim of this project.

In analogy to the thiol addition, the insertion of [1.1.1]propellane (**1**) into disulfide bonds has been known for about 30 years, but only a handful of studies looked into this reaction (Scheme 35). First mentioned and only briefly discussed was the reaction of **1** with diphenyl disulfide (**138a**) by Wiberg *et al.*, who also reacted diphenyl diselenide with **1** in the same work.[109] A more detailed description of their reaction procedure was part of another publication and revealed the initiation of the radical reaction with white light.[73] They obtained 1,3-bis(phenylthio)-bicyclo[1.1.1]pentane (**106a**) in 45% yield and showed that a conversion back to **1** was possible with lithium 4,4'-di-*tert*-butylbiphenylide (LDBB, not shown).[110]

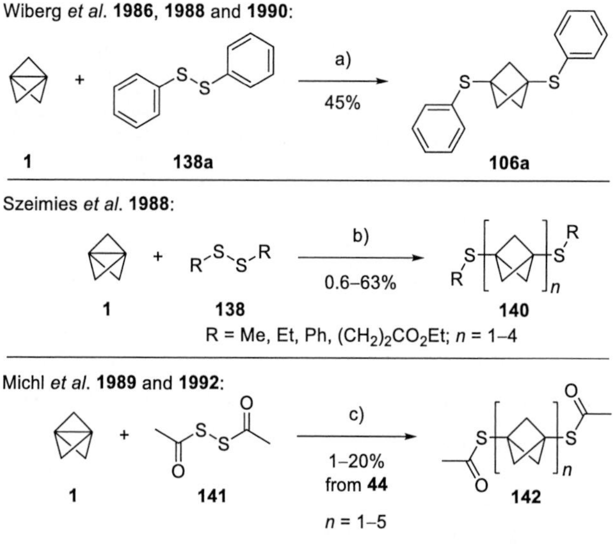

Scheme 35. Historic overview of reactions of disulfides with **1**. *a)* 1.00 equiv. **1**, 1.81 equiv. **138a**, *hv* (visible), rt, 16 h;[73, 109-110] *b)* 0.71–3.00 equiv. **1**, 1.00 equiv. **138**, 0.6–1 mol% AIBN, Et₂O, 80 °C, 4 h;[111] *c)* 3.00 equiv. **1**, 1.00 equiv. **141**, Et₂O, *hv* (UV), rt, 6 h.[112-113]

Szeimies *et al.* initiated the reaction by heating azobisisobutyronitrile (AIBN) and focused on alkyl disulfides.[111] The harsh conditions led to oligo- and polymerization. The oligomers **140** were

---

[2] Excerpts of this chapter were already published in
R. M. Bär, G. Heinrich, M. Nieger, O. Fuhr, S. Bräse, *Beilstein J. Org. Chem.* **2019**, *15*, 1172–1180.

separated and analyzed up to the corresponding [4]staffanes. Through variations in the 1:**138** ratio the yields of the different products could be influenced.

In contrast to the former two groups Michl *et al.* aimed at the synthesis of longer [*n*]staffanes (*n* up to 5) instead of the BCPs. The rigid staffane backbone should find application in a molecular construction set for nanotechnology.[112-113] In their approach the reaction of diacetyl disulfide (**141**) with **1** was initiated with UV-light and some products of different lengths (up to [5]staffanes) were hydrolyzed to the thiols and further modified.

None of the mentioned studies explored a selective synthesis of BCPs instead of mixtures with [*n*]staffanes, substituted aromatic disulfides or the possible synthesis of non-symmetrically substituted BCPs. All of these points will be addressed in this work in the following chapters.

5.2.1 Optimization

With different initiation methods reported in the literature leading to product mixtures, the aim of the optimization was to suppress oligomerization and obtain high yields of BCP. Therefore, different light sources were compared in the reaction of **1** with **138a** in a 1:1 ratio (Scheme 36). The uncommon use of ether solvents in a radical reaction was accepted, as a solvent switch in the synthesis of **1** results in a significant drop in yield.[114] Also, previous studies of this reaction were conducted in Et$_2$O without major issues.[73, 111] The reaction was monitored by GC-MS for 1 h (Figure 9). The relative conversion was determined by integration of the signals of **138a** and the products **106a** and **143a**. No other signals were observed in GC-MS and the sum of the integrals was defined as 100%. As no internal standard was used, this method does not provide absolute concentrations or yields, but rather an estimation.

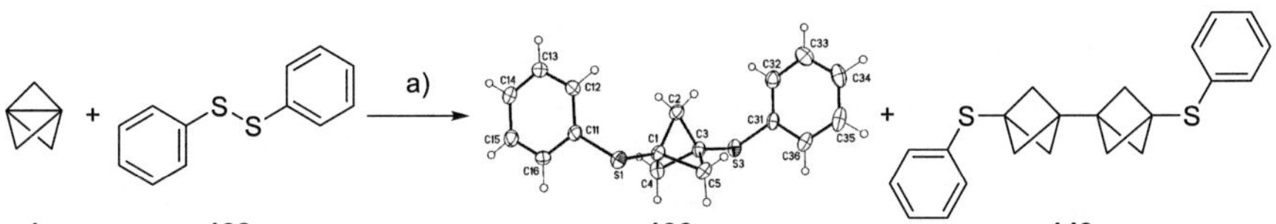

Scheme 36. Model reaction for the optimization of **1** insertion into disulfide bonds. The light source was varied and a radical initiator was used in one case (see Figure 9). *a)* 1.00 equiv. **1**, 1.00 equiv. **138a**, THF, *hv*, rt, 1 h. The structure of **106a** could be determined by single crystal X-ray diffraction (displacement parameters are drawn at 50% probability level). Crystallographic data can be found in chapter 7.4.

Stirring of the reaction mixture in darkness (green) led to no consumption of the disulfide **138a** at all. Even with exposure to daylight (yellow) only minor product formation (< 10%) was observed. The use of a 500 W halogen lamp (orange) led to almost full conversion of the starting material after 1 h. By the addition of 10 mol% of the radical initiator di-*tert*-butylperoxide (DTBP) (blue), the conversion could be accelerated. However, during the work-up of this reaction, a white, insoluble solid was discovered, presumably due to polymerization. The fastest conversion was observed when the reaction mixture was irradiated with UV-light (256 nm, 500 W) in a photoreactor with ~90% conversion after 15 min. With these promising results, an irradiation time of 20 min in the photoreactor was chosen. The absence of a radical initiator seemed to be advantageous for a selective BCP synthesis.

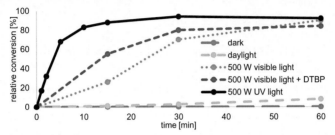

Figure 9. Relative conversion of the insertion of **1** into disulfide **138a** determined by GC-MS. Different light sources were applied. The addition of 10 mol% of radical initiator di-*tert*-butylperoxide (DTBP) led to the formation of insoluble polymer.

Szeimies *et al.* showed before that the ratio of **1:138** has a significant influence on the formation of BCP vs. [*n*]staffanes.[111] As this ratio was varied with substituted aromatic disulfides as well, the results are part of the following scope chapter.

5.2.2   Scope

The ratio of **1:138** was varied for the unsubstituted diphenyl disulfide (**138a**) and showed a smooth trend in the corresponding product ratio (Table 4). With a threefold excess of the disulfide the sole product was the BCP **106a** in 98% isolated yield (entry 1). By increasing the amount of **1** up to a ratio of 3:1 the yield of the BCP **106a** dropped to 33% and the [*2*]staffane **143a** was isolated in 20% yield (entry 4). Longer [*n*]staffanes were not observed in any of these reactions.

With practical conditions for a high-yielding BCP synthesis substituted aromatic disulfides were investigated. To obtain the corresponding [*2*]staffanes **143** as well, most of the reactions were performed with two different **1:138** ratios. The substitution with chlorides, methyl or methoxy groups did not lead to any loss in yield. In the cases of **138d–f** traces (< 5%) of the [*2*]staffanes **143d–f** could be detected, but the products **106e** and **106f** were still obtained in excellent yields (entries 10 and 12). The 2,6-dichloro substituted product **106d** and the 2-phenyl substituted product **106g** were obtained in a lower yield (entries 9 and 14). Steric hindrance could be an explanation for this. BCP **106d** could not be separated from the corresponding disulfide by chromatography, therefore the yield was determined by ^{1}H NMR spectroscopy. A ^{1}H NMR spectrum of the mixture can be found in chapter 7.3 Selected NMR spectra.

Table 4. Insertion of **1** into substituted aromatic disulfides **138**. By variation of the **1**:**138** ratio the selectivity for BCP vs [2]staffane could be influenced. *a)* **1**, **138**, THF, *hv* (256 nm, 500 W), rt, 20 min. [a] Isolated yield, purified by column chromatography; [b] not observed; [c] isolated yield, purified by preparative TLC; [d] not isolated, yield determined by ^{1}H NMR.

Entry	Disulfide	R	Ratio 1:138	Yield BCP [%]	Yield [2]staffane [%]
1	138a	H	1:3	98[a]	—[b]
2	138a	H	1:1	51[c]	5[c]
3	138a	H	2:1	32[c]	10[c]
4	138a	H	3:1	33[c]	20[c]
5	138b	4-Cl	1:3	98[a]	—[b]
6	138b	4-Cl	2:1	34[c]	15[c]
7	138c	3,5-Cl$_2$	1:3	96[a]	—[b]
8	138c	3,5-Cl$_2$	2:1	34[c]	8[c]
9	138d	2,6-Cl$_2$	1:3	32[d]	traces[d]
10	138e	4-Me	1:3	96[a]	traces[d]
11	138e	4-Me	2:1	35[c]	12[c]
12	138f	4-OMe	1:3	94[a]	traces[d]
13	138g	2-Ph	1:3	61[a]	—[b]

Unsuccessful substrates of this reaction are shown in Scheme 37. The insertion of **1** into 4-aminophenyl disulfide (**138h**) seemed successful at first (according to TLC), but the product **106h** decomposed during the purification (column chromatography on SiO$_2$). Another stationary phase like alumina oxide could possibly solve this issue, but this particular substrate was not further pursued.

The reaction of disulfide **138i** with **1** would possibly lead to the strained product **106i**. However, no conversion was observed for this substrate under the standard conditions. The formed thiophenoxyradicals presumably react back to the disulfide immediately.

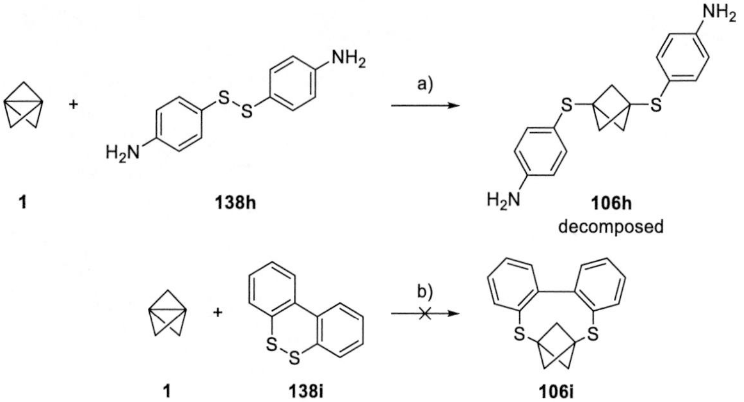

Scheme 37. Unsuccessful insertions of **1** into aromatic disulfides **138**. Product **106h** was presumably formed, but decomposed during purification. For disulfide **138i** no conversion could be observed. *a)* 1.00 equiv. **1**, 3.00 equiv. **138h**, THF, *hv* (256 nm, 500 W), rt, 20 min; *b)* 1.00 equiv. **1**, 3.00 equiv. **138i**, THF, *hv* (256 nm, 500 W), rt, 20 min.

To expand the scope towards heteroaromatic disulfides, 2,2'-dipyridyl disulfide (**144**) was reacted with **1** as a proof-of-concept (Scheme 38). The product **145** was obtained in a fair yield of 46%.

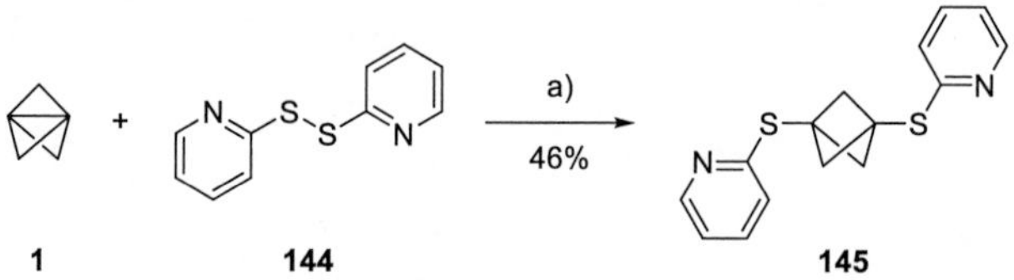

Scheme 38. Heteroaromatic disulfide **144** reacted with **1** to BCP **145** in fair yield. *a)* 1.00 equiv. **1**, 3.00 equiv. **144**, THF, *hv* (256 nm, 500 W), rt, 20 min.

As the initiation of this reaction is performed with UV-light, it is difficult to apply the method to alkyl disulfides that do not absorb at 256 nm. However, benzyl groups should be sufficient to absorb the light and initiate the homolytic cleavage. Therefore, dibenzyl disulfide (**146**) was used in the reaction and BCP **147** was successfully obtained (Scheme 39). In an attempt to obtain the [2]staffane **148** as well, a mixture of products was obtained that could not be separated by chromatography. Interestingly, **148** formed single crystals from this mixture that were suitable for X-Ray diffraction. The rod-like structure of **148** makes it easy to understand Michl's intention to use staffanes as a stick in nanotechnology.

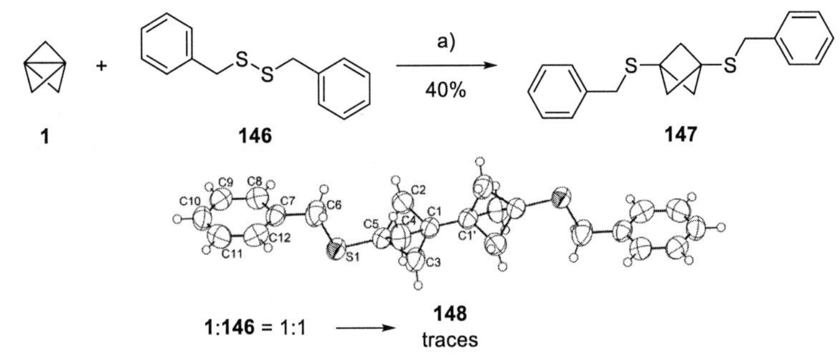

Scheme 39. Insertion of 1 into dibenzyl disulfide (146). With a 1:3 ratio of 1:146 BCP 147 was the only product. With a 1:1 ratio a mixture of 147 and [2]staffane 148 was obtained and single crystals of 148 could be analyzed by X-Ray diffraction (displacement parameters are drawn at 50% probability level). Crystallographic data can be found in chapter 7.4. *a)* 1.00 equiv. 1, 3.00 equiv. 146, THF, *hv* (256 nm, 500 W), rt, 20 min.

The molecular structure of **148** allows comparison to **106a** and reveals slightly longer bond distances between the bridgehead carbon atoms (C1–C3 = 1.844 Å for **106a**; C1–C5 = 1.857 Å for **148**). As the second BCP is an EDG the s character of the hybrid atom orbital for the exocyclic bond is decreased and therefore the bond length is increased (Bent's rule).[115] This observation is in accordance with the literature.[116]

5.2.3   Mechanistic considerations

The reaction of disulfides **138** with [1.1.1]propellane (**1**) has been described as free-radical addition already.[73, 109, 111-112] The proposed mechanism in Scheme 40 shows the homolytic cleavage of the disulfide by light and the subsequent reaction with **1** to form a carbon-centered radical. In a chain propagation the reaction with another equivalent of disulfide **138** releases the desired BCP product **106**. The [2]staffane **143** can be obtained either through dimerization of the BCP radical or through the reaction with a second equivalent of **1** and subsequent chain propagation.

Based on this mechanism, the yield of **143** and longer staffanes should increase with the concentration of **1**. The dimerization should be less dominant as the concentration of the BCP radical is presumably low. From the synthetically obtained results it seems as the formation of staffanes is less favored than the propagation with the disulfide. This observation is in accordance with previous calculations and experimental results.[81, 86]

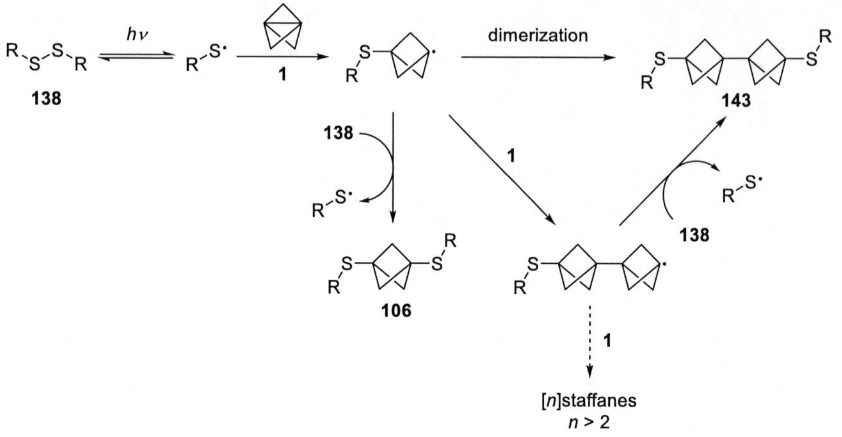

Scheme 40. Proposed mechanism of the insertion of **1** into disulfide bonds. The mechanism is based on a free radical addition as stated by others before.[73, 109, 111-112]

5.2.4 Reactions with two disulfides

In the former chapter the free radical addition of thiyl radicals to **1** with a BCP radical intermediate was shown (Scheme 40). Similar BCP radical intermediates have been trapped with other reagents to obtain non-symmetrically substituted BCPs.[81, 86] This suggests the synthesis of such a BCP from a 1:1 mixture of disulfides. If the reaction rate doesn't differ for the different disulfides a statistical mixture should be obtained with the non-symmetrically substituted BCP as the major product (50%).

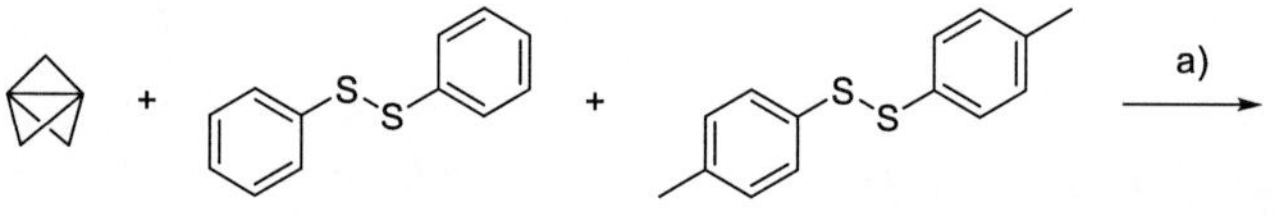

1 **138a** **138e**

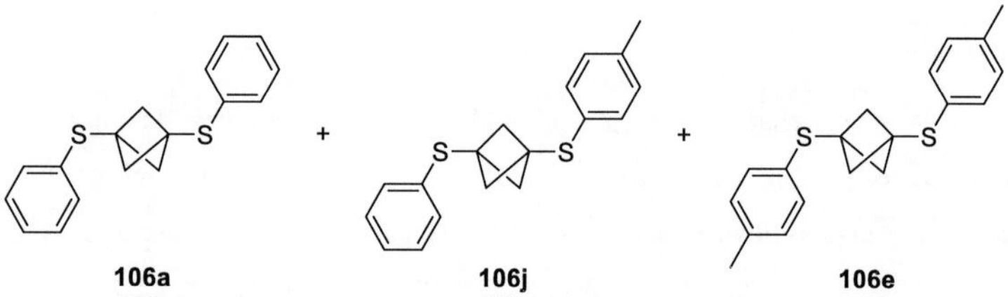

106a 18% **106j** 45% **106e** 24%

Scheme 41. The reaction of **1** with two disulfides **138a** and **138e** led to the expected product mixture with non-symmetrically substituted BCP **106j** as the major product. The product mixture was not separable by column chromatography, the yield was determined by ^{1}H NMR. The spectrum of the resulting mixture can be found in chapter 7.3 Selected NMR spectra. *a)* 1.00 equiv. **1**, 1.50 equiv. **138a**, 1.50 equiv. **138d**, THF, *hν* (256 nm, 500 W), rt, 20 min.

In order to investigate this novel approach to non-symmetrically substituted BCPs the disulfides **138a** and **138e** were premixed and reacted with **1** under standard conditions (Scheme 41). The obtained product mixture could not be separated, but the yield could be determined by ^{1}H NMR (see ^{1}H NMR in chapter 7.3). To facilitate the separation of the products by column chromatography and/or preparative TLC **138e** was exchanged for the slightly more polar disulfide **138f** (Scheme 42). The excess of disulfides was removed by column chromatography and the product mixture could be separated by preparative TLC. The obtained yields are almost in accordance with the expectations, with a minor preference towards **138f**. As mentioned in chapter 5.1.3 the 4-methoxythiophenoxy radical seems to be formed and/or react faster with **1** than the corresponding radical from **138a**. The calculated bond-dissociation energy (BDE) of **138f** can be found in the literature and is lower than the BDE of **138a** (49.0 kcal/mol vs. 54.5 kcal/mol).[117] This difference suggests a faster formation of the 4-methoxythiophenoxy radical.

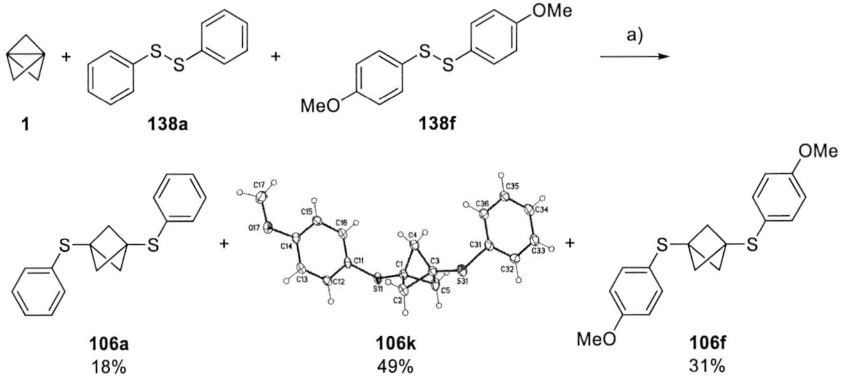

Scheme 42. Reaction of disulfides **138a** and **138f** with **1** led to a mixture of products that could be separated by preparative TLC. The non-symmetrically substituted BCP **106k** could be crystallized and analyzed by X-Ray diffraction (displacement parameters are drawn at 50% probability level). Crystallographic data can be found in chapter 7.4. *a)* 1.00 equiv. **1**, 1.50 equiv. **138a**, 1.50 equiv. **138f**, THF, *hν* (256 nm, 500 W), rt, 20 min.

It is noteworthy, that this reaction does not require the synthesis of a non-symmetric disulfide to obtain a non-symmetrically substituted product. However, it is also possible to start from non-symmetric disulfides as Lukas Langer (Bräse group) showed in his master thesis.[118]

The insertion of **1** into disulfides was applied to different aromatic disulfides **138**, a heteroaromatic disulfide **144** and benzyl disulfide **146**. The limitation of the developed method to aromatic residues lies in the initiation with UV-light. The use of two symmetrical disulfides to obtain non-symmetrically substituted BCPs is of great interest for applications in bioisosteres and bioconjugations. The extension of this method to 2-substituted [1.1.1]propellanes would be very interesting for the synthesis of ADCs (see chapter 3.2.1). The BCP could work as a stable linker between the naturally occurring disulfides in the antibody and the active drug compound.

5.3 Modifications of bicyclo[1.1.1]pentylsulfides

For possible applications in medicinal chemistry or related fields, different aspects need to be addressed concerning the versatility of BCP sulfides. Although sulfides appear as a motif in some bioactive compounds, the majority of sulfur-containing drugs is based on more polar derivatives.[119] A straightforward way to achieve this increase in polarity is the oxidation to sulfoxides and sulfones. Also imination of BCP sulfides or sulfoxides can lead to interesting scaffolds, e.g. sulfoximines. The presence of a polar sulfur-based functional group can be used as a directing group for further aromatic substitutions. Reactions like these will be described in this chapter.

5.3.1 Bicyclo[1.1.1]pentylsulfoxides[3]

The sulfone **149a** was already synthesized during my master thesis by oxidation with an excess of *m*-chloroperbenzoic acid (*m*CPBA).[96] To obtain the sulfoxide **107a** in a high yield, the oxidation was performed with increasing amounts of *m*CPBA and analyzed by HPLC (Scheme 43 and Figure 10). Therefore, the absorption signals of **104a** (black), **107a** (blue) and **149a** (orange) at 256 nm were integrated and the sum of the integrals was defined as 100%. As no internal standard was used, this method does not provide absolute concentrations or yields, but rather an estimation.

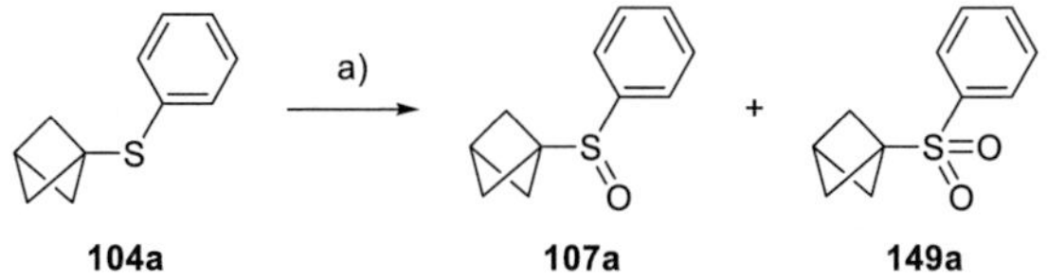

Scheme 43. Oxidation of sulfide **104a** with increasing amounts of *m*CPBA. The reactions were analyzed by HPLC (Figure 10). *a)* 1.00 equiv. **104a**, 0.00–3.08 equiv. *m*CPBA, CH$_2$Cl$_2$, rt, 5 min.

The formation of sulfoxide **107a** seems to be most selective with 1.00–1.50 equiv. of *m*CPBA and in a larger scale (500 mg starting material) **107a** could be isolated in 67% yield with 1.15 equiv. of the oxidant (Scheme 44).

[3] Excerpts of this chapter were already published in
R. M. Bär, S. Kirschner, M. Nieger, S. Bräse, *Chem. Eur. J.* **2018**, *24*, 1373–1382.

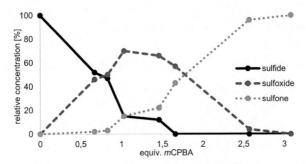

Figure 10. Relative concentration of sulfide **104a**, sulfoxide **107a** and sulfone **149a** in the oxidation reaction shown in Scheme 43, determined by HPLC. The amount of oxidant mCPBA was varied in different approaches.

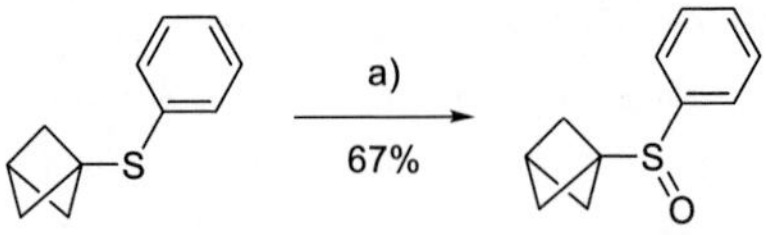

Scheme 44. Scale-up of the sulfoxide synthesis with 500 mg (2.84 mmol) starting material. *a)* 1.00 equiv. **104a**, 1.15 equiv. mCPBA, CH$_2$Cl$_2$, rt, 5 min.

An enantioselective synthesis of **107a** was not tested in this work. There are established methods for similar substrates with chiral titanium[120] or aluminum complexes[121] that could be part of a subsequent study.

5.3.2   Bicyclo[1.1.1]pentylsulfone as direct metalation group

To obtain polysubstituted aromatics or to modify a compound in a late-stage direct metalation groups (DMGs) can be useful tools for *ortho*-substitutions.[122] Iwao, Snieckus *et al.* demonstrated that *tert*-butylsulfone can be an effective DMG in lithiations (Scheme 45).[123] The subsequent reaction with electrophiles can be used to access a variety of products.

Iwao, Snieckus *et al.* **1989:**

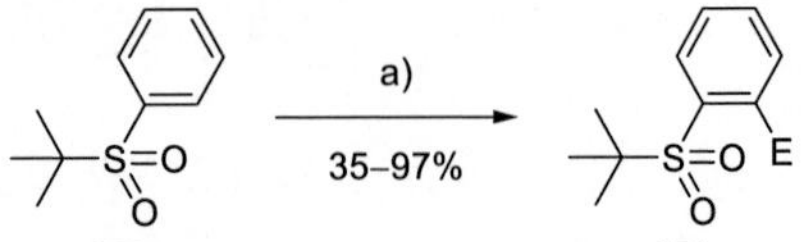

Scheme 45. Investigation of *tert*-butylsulfone as direct metalation group.[123] *a)* 1.00 equiv. **150**, 1.00 equiv. RLi, THF, –78 °C, 30 min, then 2.00 equiv. electrophile.

As the BCP sulfone **149a** was readily available[96] and shares a great similarity with *tert*-butylsulfone **150**, the question was raised whether similar reactions would be possible with

149a. If the BCP sulfone would not direct the lithiation, a deprotonation of the terminal BCP would also be possible with a strong base. Due to the increased s character of strained alkanes the acidity of the exocyclic C–H is increased.[124] Literature values for the s character of unsubstituted BCP range from 32.6–34%.[125-126] For sulfone **149a** a s character of 30.2% was determined from the $^1J_{CH}$-coupling constant (151 Hz).[127-128] It should be noted, that the accuracy of the s character calculation from the $^1J_{CH}$-coupling constant is lower when EWG, like the sulfone, are attached.

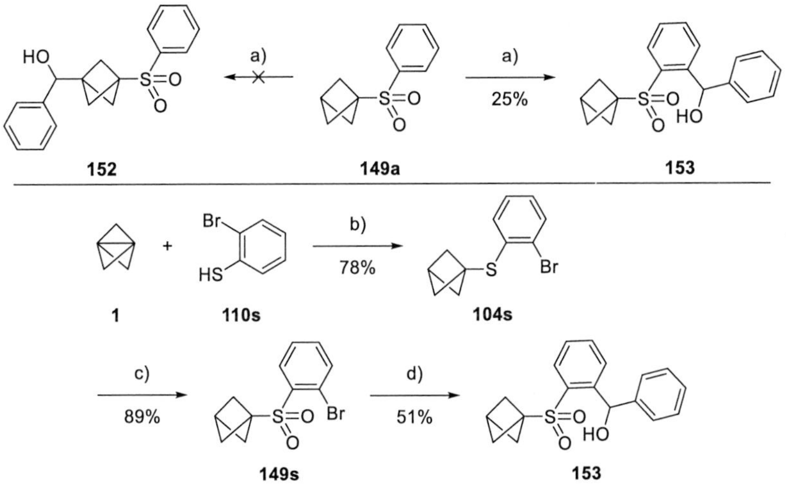

Scheme 46. Lithiation of **149a** takes place in *ortho*-position of the phenyl substituent. To confirm the regiochemistry of the product **153**, *ortho*-bromo substituted sulfone **149s** was substituted to give the same product **153**. *a)* 1.00 equiv. **149a**, 1.05 equiv. *t*-BuLi, THF, –78 °C, 1 h, then 1.51 equiv. PhCHO, –78 °C to rt, 1 h; *b)* 1.00 equiv. **1**, 1.66 equiv. **110s**, rt, 30 min; *c)* 1.00 equiv. **104s**, 3.01 equiv. *m*CPBA, CH$_2$Cl$_2$, 0 °C to rt, 1 h; *d)* 1.00 equiv. **149s**, 1.06 equiv. *n*-BuLi, –78 °C, 1 h, then 1.53 equiv. PhCHO, –78 °C to rt, 1 h.

To my current knowledge, there is no method available to modify the CH of a terminal BCP directly.[4] Therefore, either outcome of the reaction would be interesting and the lithiation of **149a** was performed with *tert*-butyllithium (Scheme 46). After addition of benzaldehyde, alcohol **153** was obtained in 25% yield. Some starting material was left (not isolated) and no other major product could be detected by TLC. Presumably, the reaction conditions could be optimized to increase the yield, but this was not further pursued. The structure elucidation of **153** with the *ortho*-substitution was based on NMR spectroscopy. To confirm the structure *ortho*-bromo substituted BCP sulfone **149s** was prepared *via* the thiol addition to **1** and subsequent oxidation of **104s** with *m*CPBA. Through metal-halogen exchange with *n*-butyllithium and addition of

---

[4] After the submission of this thesis a method for the CH insertion at the terminal position of BCPs was published.[129]

benzaldehyde, product **153** could be obtained in 51% yield. The comparison of the obtained NMR spectra confirmed the identity of **153** obtained from **149a**. The ^{1}H and ^{13}C NMR spectra can be found in chapter 7.3.

5.3.3   Bicyclo[1.1.1]pentylsulfoximines[5]

Sulfoximines have been described as 'neglected opportunity' in medicinal chemistry.[130-131] The replacement of other sulfur-based functional groups by sulfoximines can have positive effects on physicochemical properties and metabolic stability.[132] Therefore, this polar sulfur(VI) group is used regularly in SAR studies and some drug candidates with sulfoximines have already been published (Figure 11).

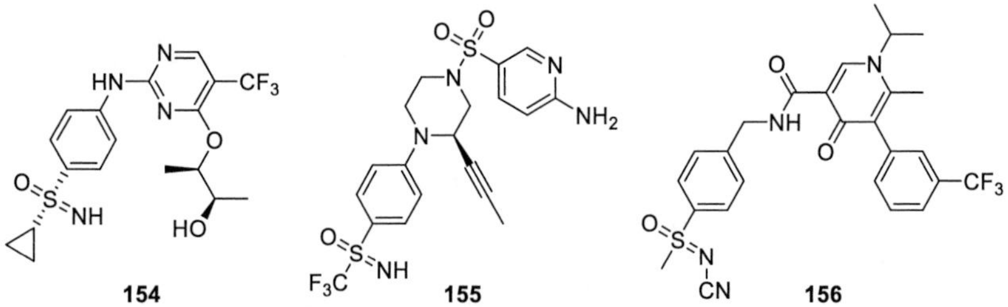

Figure 11. Examples for drug candidates containing a sulfoximine moiety. Roniciclib (**154**) was a candidate for CDK inhibition by Bayer,[133-134] compound **155** by Amgen is currently under investigation in a diabetes-related study[135] and **156** has been disclosed by Boehringer Ingelheim as inhibitor of neutrophile elastase activity.[136]

Sulfoximines can generally be accessed from the corresponding sulfides, sulfoxides or sulfilimines.[137] As both BCP sulfides and BCP sulfoxides were available by the previously described methods, the compounds **104a** and **107a** were converted to BCP sulfoximine **108a** (Scheme 47).

With conditions by Li *et al.* **108a** could be obtained in 55% yield from sulfide **104a**.[138] In a scale-up (1.20 g starting material) with increased equivalents of the reagents the yield could be slightly improved to 57%. The sulfoxide **107a** could be iminated with conditions by Luisi, Bull *et al.* to **108a** in 88% yield.[139] So with a two-step yield of 59% (from **104a**) the synthetic route through the sulfoxide results in a similar yield. As this route requires two chromatographic purification steps, the direct route from the sulfide was chosen for further substrates.

---

[5] Excerpts of this chapter were already published in
R. M. Bär, L. Langer, M. Nieger, S. Bräse, *Adv. Synth. Catal.* **2020**, *362*, 1356–1361.

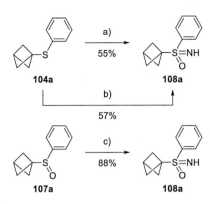

Scheme 47. Synthesis of BCP sulfoximine **108a** from the sulfide **104a** and the sulfoxide **107a**. *a)* 1.00 equiv. **104a**, 1.50 equiv. (NH₄)₂CO₃, 2.30 equiv. PhI(OAc)₂, MeOH, rt, 30 min; *b)* Scale-up with 1.20 g (6.81 mmol) starting material, 1.00 equiv. **104a**, 2.00 equiv. (NH₄)₂CO₃, 3.00 equiv. PhI(OAc)₂, MeOH, rt, 30 min; *c)* 1.00 equiv. **107a**, 1.50 equiv. H₂NCO₂NH₄, 3.00 equiv. PhI(OAc)₂, MeOH, rt, 30 min.

The reaction conditions with increased amounts of the ammonium source (2.00 equiv.) and the oxidant (3.00 equiv.) were applied to several BCP sulfides from the previous chapters (Table 5). *Para*-substituted aryl BCP sulfides were suitable substrates for this reaction. The chlorinated product **108b** was obtained in a lower yield of 34% (entry 1), whereas the tolyl derivative **108c** was obtained in a higher yield of 60% (entry 3) compared to the parent compound **108a**. This suggests a trend towards higher yields with higher electron densities in the aryl substituent. As the sulfur needs to be oxidized in the reaction this observation is in accordance with the expectations. In a reaction from sulfoxide **107b** (for the synthesis see chapter 5.4.2) to **108b** an excellent yield of 97% could be obtained (entry 2). This shows that, under the given reaction conditions, the oxidation is the limiting step in this reaction. For substrate **104d** the +M effect outweighs the –I effect of the methoxy substituent with a very good yield of 83%.

Purely alkyl substituted sulfoximines were generally obtained in lower yields. The volatile butyl substituted sulfide **105c** was converted to the sulfoximine **108e** in 43% yield (entry 5), the benzylated product **105f** was obtained in a similar yield of 41% (entry 6) and product **108g** was obtained in a poor yield of 27% (entry 7).

Table 5. Synthesis of BCP sulfoximines **108** from sulfides **104/105**. *a)* 1.00 equiv. **104/105**, 2.00 equiv. $(NH_4)_2CO_3$, 3.00 equiv. PhI(OAc)$_2$, MeOH, rt, 30 min. [a] Isolated yield; [b] synthesized from sulfoxide **107b**: 1.00 equiv. **107b**, 3.98 equiv. $(NH_4)_2CO_3$, 2.00 equiv. PhI(OAc)$_2$, MeOH, rt, 30 min.

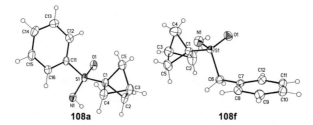

Entry	Sulfide	R	Product	Yield [%][a]
1	104b	Cl-phenyl	108b	34
2	sulfoxide 107b	Cl-phenyl	108b	97[b]
3	104d	Me-phenyl	108c	60
4	104f	OMe-phenyl	108d	83
5	105c	*n*-Bu	108e	43
6	105f	Bn	108f	41
7	105k	$(CH_2)_2CO_2Me$	108g	27

For the two products **108a** and **108f** single crystals could be obtained and the structure could be determined by X-ray diffraction (Figure 12). All of the BCP sulfoximines were synthesized in a racemic fashion and **108a** crystallized as a racemic twin.

Figure 12. Structures of **108a** and **108f** determined by single-crystal X-ray diffraction. Displacement parameters are drawn at 50% probability level. Crystallographic data can be found in chapter 7.4.

This scope investigation was conducted further in the master thesis of Lukas Langer (Bräse group).[118] The results are summarized in Table 6 to give a complete picture of this project. The key observations were a lower yield for electron-deficient aryl substituted BCPs, which is in accordance with the previous results and no conversion for *ortho*-substituted BCP sulfides due to steric hindrance. The two bis(sulfoximines) **108n** and **108o** should be obtained as two diastereomers, but could unfortunately not be separated or distinguished.

Table 6. Extension of the scope investigation for the sulfoximine synthesis (only successful substrates shown) conducted by Lukas Langer (Bräse group) during his master thesis.[118] *a)* 1.00 equiv. **104/105**, 1.50 equiv. (NH₄)₂CO₃, 2.30 equiv. PhI(OAc)₂, MeOH, rt, 30 min. [a] Isolated yield; [b] Reaction done with 4.00 equiv. (NH₄)₂CO₃ and 6.00 equiv. PhI(OAc)₂.

Entry	Sulfide	R	Product	Yield [%][a]
1	104e	*t*-Bu (4-substituted phenyl)	108h	86
2	104g	NO₂ (4-substituted phenyl)	108i	45
3	104i	Cl (3-substituted phenyl)	108j	27
4	104s	CF₃, CF₃ (3,5-substituted phenyl)	108k	39
5	105d	*t*-Bu	108l	14
6	105g	(CH₂)₂OH	108m	37
7	106a		108n	74[b]
8	157		108o	17[b]

Sulfoximines offer a handle for further modifications. As an example the *N*-arylation of **108a** was chosen and investigated. For the *N*-arylation of sulfoximines several methods are available.[140] Most commonly used are palladium-catalyzed couplings with aryl halides or sulfonates. But also iron and especially copper are suitable catalysts and have been used in combination with different coupling partners. Copper was chosen for this approach as cost effective catalyst.

Table 7. Optimization of the copper(I) catalyst for the *N*-arylation of sulfoximine **108a**. *a)* 1.00 equiv. **108a**, 1.52 equiv. **158**, *catalyst*, 2.00 equiv. Cs₂CO₃, DMF, 100 °C, 18 h. [a] Isolated yield; [b] not observed.

Entry	Catalyst loading	Catalyst	Yield [%][a]
1	10 mol%	Cu₂O	37
2	10 mol%	CuI	3
3	10 mol%	CuBr	15
4	10 mol%	CuCl	7
5	10 mol%	CuCN	15
6	10 mol%	CuTC	1
7	10 mol%	CuOTf-Toluene	15
8	10 mol%	Cu(CH₃CN)₄PF₆	31
9	10 mol%	CuOAc	39
10	20 mol%	Cu₂O	42
11	30 mol%	CuOAc	54
12	10 mol%	CuO	–[b]

Starting with literature based ligand-free conditions,[141] the copper(I) source was varied to optimize the reaction (Table 7). The initial yield with copper(I) oxide (37%, entry 1) could not be outperformed with copper(I) halides (entries 2–4) or copper(I) cyanide (entry 5). More soluble copper(I) salts and complexes did not improve the yield either (entries 6–8). Only copper(I) acetate led to a similar yield of 39% (entry 9). A higher catalyst loading did only improve the yield in small steps (entries 10–11), but led to the highest yield of 54% in this screening with 30 mol% CuOAc. For further optimization copper(I) oxide was chosen due to lower costs of the catalyst. To ensure the dependency of this reaction on copper(I), copper(II) oxide was used as well, but did not lead to any product formation (entry 12).

The optimization of the *N*-arylation and a subsequent scope investigation was continued by Lukas Langer (Bräse group) in the course of his master thesis.[118] Key steps in the optimization were the exchange of Cs₂CO₃ to KO*t*-Bu and the addition of *N*,*N* '-dimethylethylenediamine (DMEDA) as a ligand. With an optimized yield of 70% for model compound **108a** the scope was investigated (Table 8). Substitutions in *meta-*, *ortho-* or *para*-position did not shut down the reaction and a very good functional group tolerance was observed. Only carboxylic esters were not tolerated as they were presumably saponified to the potassium carboxylate (not shown). Bromoiodobenzenes led to product mixtures that could not be separated (Entries 18–20).

Table 8. Scope investigations of the *N*-arylation of BCP sulfoximines **108** (only successful substrates shown) conducted by Lukas Langer (Bräse group) during his master thesis.[118] *a)* 1.00 equiv. **108**, 1.50 equiv. **160**, 20 mol% Cu$_2$O, 20 mol% DMEDA, 2.00 equiv. KO*t*-Bu, DMF, 100 °C, 18 h. [a] Isolated yield; [b] product mixtures obtained, yield determined by ^{1}H NMR spectroscopy.

Entry	Sulfoximine	R	Ar-X	Product	Yield [%][a]
1	108a	Ph	PhI	159a	70
2	108a	Ph		159b	60
3	108a	Ph	OMe	159c	57
4	108a	Ph	Cl	159d	48
5	108a	Ph	NO$_2$	159e	42
6	108a	Ph		159f	56
7	108a	Ph		159g	75
8	108a	Ph	OMe	159h	70
9	108a	Ph	Cl	159i	69
10	108a	Ph		159j	46
11	108a	Ph		159k	70
12	108a	Ph	MeO	159l	41

Entry	Sulfoximine	R	Ar-X	Product	Yield [%][a]
13	108a	Ph		159m	26
14	108a	Ph		159n	33
15	108a	Ph		159o	35
16	108f	Bn		159p	68
17	108e	n-Bu		159q	46
18	108a	Ph		159r (X = Br) 22%[b] 159s (X = I) 18%[b]	
19	108a	Ph		159t (X = Br) 26%[b] 159u (X = I) 26%[b]	
20	108a	Ph		159v (X = Br) 17%[b] 159w (X = I) 17%[b]	

The different modifications of BCP sulfides shown in this chapter highlight the versatility of the compounds. Sulfur-based functional groups can easily be oxidized and iminated to obtain more polar and drug-like products. It was shown that BCP sulfone can act as a directing group in lithiation reactions. Sulfoximines further offer the free *NH* group for substitutions. The *N*-arylation was optimized and the scope was investigated together with Lukas Langer.

5.4   Sodium bicyclo[1.1.1]pentanesulfinate[6]

All previously discussed BCP syntheses require the use of the volatile [1.1.1]propellane (**1**). For some research groups, e.g. in pharmaceutical industry, the synthesis and handling of this precursor involves some unwanted precautions like Schlenk techniques or the use of large amounts of organolithium reagents. Readily available, bench-stable precursors for BCPs are therefore highly desirable (Scheme 48). The most commonly used precursor in this context is 1,3-diacetylbicyclo[1.1.1]pentane (**53**), which can be converted to the non-symmetrical building block **55** in several steps (see chapter 3.3.2).[77] There have been great efforts to optimize the synthesis of BCP amine (**76**, shown as hydrochloride)[66, 84] and substituted derivatives **82**.[86-87] A recently published synthesis of silaborated BCP **160** enables the modular synthesis of non-symmetrically substituted BCPs through modification of the boronic ester and subsequent oxidative cleavage of the silyl group.[142]

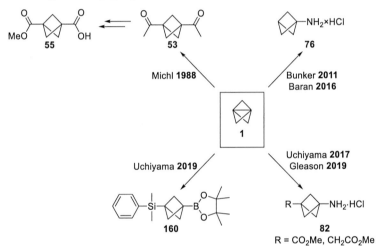

Scheme 48. Examples for bench-stable BCP building blocks derived from **1**. References: Michl, **1988**;[77] Bunker **2011**;[84] Baran **2016**;[66] Uchiyama **2017**;[86] Gleason **2019**;[87] Uchiyama **2019**.[142]

As this thesis focusses on the synthesis of sulfur-containing BCPs, a bench-stable precursor for such compounds was a major aim. In the course of this project the so-far unprecedented synthesis of BCP sulfonamides should be tackled as well. Sulfonamides possess biological activity in many cases and are especially known for their antimicrobial properties.[143] The standard procedure for

---

[6] Excerpts of this chapter were already published in
R. M. Bär, P. J. Gross, M. Nieger, S. Bräse, *Chem. Eur. J.* **2020**, *26*, 4242–4245.

their synthesis is the nucleophilic substitution of sulfonyl chlorides with amines (Scheme 49).[144] Attempts to synthesize BCP sulfonyl chloride through deprotection and oxidation/chlorination of BCP thiol (**135**, see chapter 5.1.2) were unsuccessful. An alternative strategy was pursued, based on work from Baskin and Wang, who used the sulfinate intermediate in the synthesis of primary sulfonamides.[145]

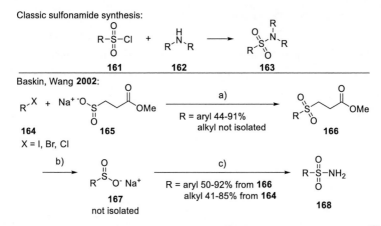

Scheme 49. Sulfonamide synthesis *via* the sulfonyl chloride **161** or *via* a sulfinate intermediate **167**.[145] *a)* R = aryl: 1.00 equiv. **164**, 3.00 equiv. **165**, 3.00 equiv. CuI, DMSO, 110 °C, 3–24 h; R = alkyl: 1.00 equiv. **164**, 1.20 equiv. **165**, DMSO, rt, 10 min to 24 h; *b)* 1.00 equiv. **166**, 1.00 equiv. NaOMe, DMSO, rt, 10 min; *c)* crude **167** in DMSO, 5.00 equiv. NH$_2$OSO$_3$H, 3.81–3.93 equiv. NaOAc, H$_2$O, rt, 20 h.

5.4.1   Synthesis and scale-up

BCP sulfide **105k** is a suitable precursor for the sulfinate synthesis and was already synthesized through the thiol addition (see chapter 5.1.2). The sulfide was obtained in a quantitative fashion from **1** on a small scale (440 µmol) and in order to scale-up the reaction the thiol **132k** was added directly to a freshly made solution of **1** (Scheme 50). After the standard work-up, **105k** was obtained in 79% over two steps on a 74 mmol scale. The oxidation to sulfone **149k** could be achieved with *m*CPBA in a short reaction time and with a very good yield of 82%. Although the reaction mixture was washed with NaOH solution, traces of *m*-chlorobenzoic acid could be found in the final product **169**. Therefore, an alternative oxidation procedure was applied by using oxone. The slightly decreased yield of 72% was accepted as it resulted in a product with higher purity. In analogy to the sulfinate precursor **166**, BCP sulfone **149k** could be converted to BCP sulfinate **169** in a retro-Michael reaction. The final product sodium bicyclo[1.1.1]pentanesulfinate (BCP-SO$_2$Na, **169**) was obtained in quantitative yield on a 61 mmol scale (9.4 g).

The four step synthetic route could be performed without any chromatography or (re-)crystallization. A further scale-up in an industrial production facility should be possible based on these results. It should be noted, that more cost effective oxidations of sulfides are known using hydrogen peroxide and a metal-based catalyst, e.g. tungstates.[146] They were not part of this study, but should be considered in further investigations. Also, the released acrylate in the last step could be condensed and recycled in an industrial process.

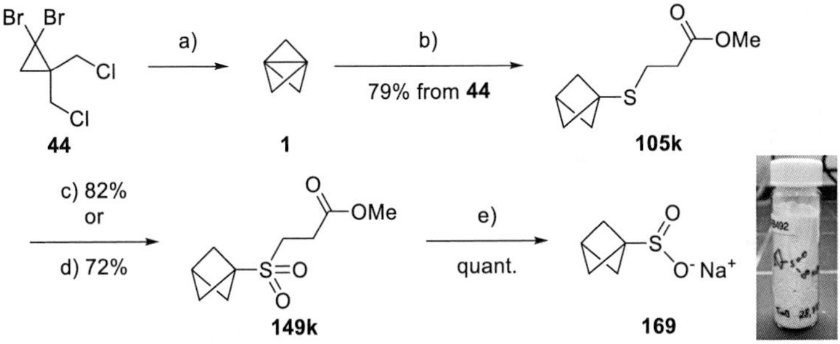

Scheme 50. Synthesis of BCP-SO$_2$Na (**169**) through retro-Michael reaction of sulfone **149k**. *a)* 1.00 equiv. **44**, 2.02 equiv. PhLi, Et$_2$O, −78 to 0 °C, 2 h; *b)* **1** in Et$_2$O, 1.30 equiv. methyl 3-mercaptopropionate (**132k**), rt, 30 min; *c)* 1.00 equiv. **105k**, 2.62 equiv. *m*CPBA, CH$_2$Cl$_2$, 0 °C to rt, 1 h; *d)* 1.00 equiv. **105k**, 2.00 equiv. oxone, 1,4-dioxane/water, rt, 18 h; *e)* 1.00 equiv. **149k**, 1.00 equiv. NaOMe, THF, rt, 20 min.

BCP-SO$_2$Na (**169**) could now be accessed in large quantity and good purity and applications of the novel building block will be discussed in the following chapter.

5.4.2 Application as building block

Sulfinates offer a variety of synthetic applications with the sulfur and the oxygen as two nucleophilic centers.[147] Most commonly the salts are deployed in the synthesis of sulfones either through nucleophilic substitution or metal-catalyzed couplings. Halogenation inverts the reactivity and enables the synthesis of sulfonic esters, sulfonamides and others.

The nucleophilic aromatic substitution (S$_N$Ar) of electron deficient aryl halides was investigated first (Table 9). Aryl fluorides **170a–c** could be substituted in good to excellent yields (entries 1–3). The nitro group offers access to valuable amines by reduction. The obtained products **149b–d** could in theory be obtained through the thiol addition and subsequent oxidation as well. But for heteroaromatic thiols the addition to **1** was not successful (see chapter 5.1.2). Hence, the reaction of **169** with quinoline **170d** (entry 4) and isoquinoline **170e** (entry 5) enables access to heteroaryl BCP sulfones for the first time.

For several products of the S$_N$Ar reaction single crystals could be obtained and the structure could be determined by X-ray diffraction (Figure 13). The motivation for the synthesis of novel BCP building blocks is the similarity in spatial features with benzenes that enable a bioisosteric replacement with the strained hydrocarbon (see chapter 3.1.3). At this point, the opportunity was taken and an analogues sulfone to **149b** was synthesized from sodium benzenesulfinate (**171**, Scheme 51). This product could be analyzed by single crystal X-ray diffraction as well and the size and angle of the terminal BCP could be compared with the phenyl substituent (Figure 14). The similar length and angle of the BCP and the phenyl substituent become visible in this presentation. But there are still small differences that can be important for the application of this motif in SAR studies. The distance in **172** between the sulfur atom and the carbon atom in *para*-position is 4.51 Å, whereas the analogue distance in the BCP compound **149b** is only 3.60 Å. The angle also differs slightly from the desired 180 ° (179.7 ° for the phenyl) with 177.9 ° for the BCP.

Table 9. Nucleophilic aromatic substitution of electron deficient aryl halides **170**. *a)* for X = F: 1.30 equiv. **169**, 1.00 equiv. **170**, DMF, 80–100 °C, 16–72 h; *b)* for X = Cl: 1.30 equiv. **169**, 1.00 equiv. **170**, 1.50 equiv. K$_2$CO$_3$, 120 °C, 16 h.

Entry	Ar-X	Product	Yield [%][a]
1	**170a**	**149b**	96
2	**170b**	**149c**	98
3	**170c**	**149d**	75
4	**170d**	**149e**	29
5	**170e**	**149f**	41

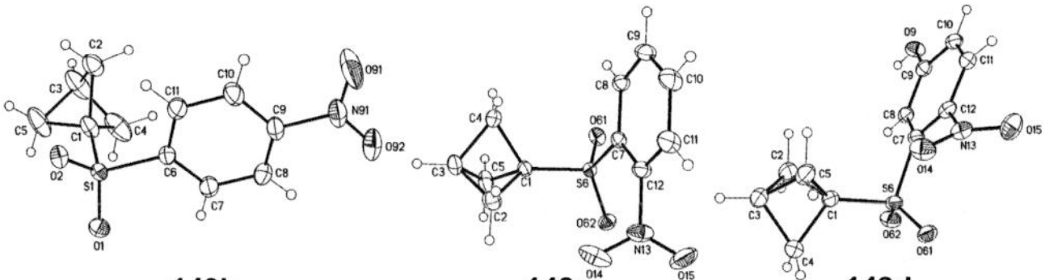

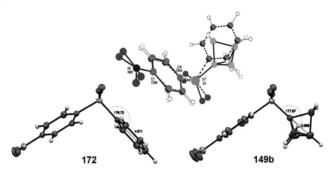

149b   **149c**   **149d**

Figure 13. Structures of BCP sulfones **149b–d** determined by single-crystal X-ray diffraction. Displacement parameters are drawn at 50% probability level. Crystallographic data can be found in chapter 7.4.

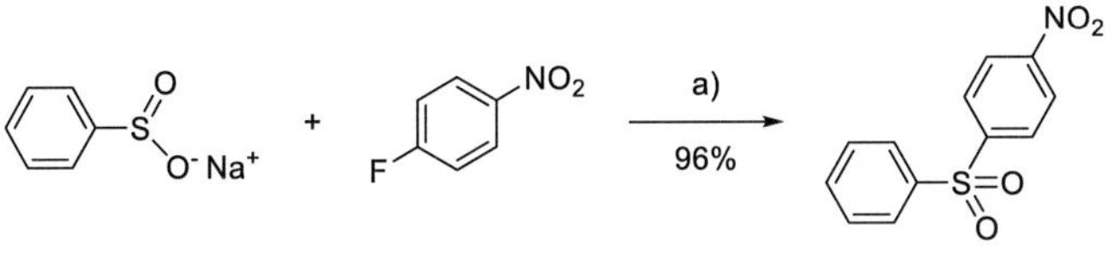

Scheme 51. Nucleophilic aromatic substitution of aryl fluoride **170a** with benzenesulfinate **171**. *a)* 1.30 equiv. **171**, 1.00 equiv. **170a**, DMF, 80 °C, 2 h.

172   **149b**

Figure 14. Comparison of the molecular structures of **172** and **149b**. The angles and distances between the sulfur atom of the sulfone and the carbon in *para*-position for **172** and 3-position for the BCP **149b** are indicated.

For the synthesis of more electron rich aryl sulfones, metal-catalyzed couplings can be applied. As an example sulfinate **169** was reacted with aryl bromide **170f** in a copper-catalyzed reaction and **149g** was obtained in 73% yield (Scheme 52). The same reaction conditions led to a poor yield of 18% for the *ortho*-substituted sulfone **149s**. The sulfone **149s** was previously obtained in chapter 5.3.2 through a two step procedure from **1**. The lower yield compared to **149s** can be explained by two factors. First, the two halides compete in this reaction and only **149s** was isolated. A double substitution could also be possible. Second, the *ortho*-substituent leads to steric hindrance.

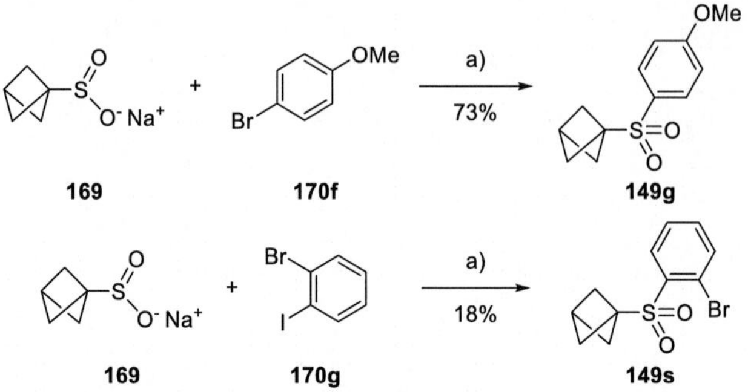

Scheme 52. Copper-catalyzed Suzuki-Miyaura type reaction between sulfinate **169** and aryl halides. *a)* 1.30 equiv. **169**, 1.00 equiv. **170f**, 10 mol% CuI, 20 mol% L-proline, 1.00 equiv. K$_2$CO$_3$, DMSO, 110 °C, 22 h; *b)* 1.31 equiv. **169**, 1.00 equiv. **170g**, 10 mol% CuI, 20 mol% L-proline, 1.00 equiv. K$_2$CO$_3$, DMSO, 100 °C, 18 h.

Alkyl substituted sulfones can be accessed by nucleophilic substitution of alkyl halides with sulfinates.[147] The smallest example of these compounds was synthesized from **169** and iodomethane (**173**) and the structure of **149h** could be determined by X-ray diffraction (Scheme 53).

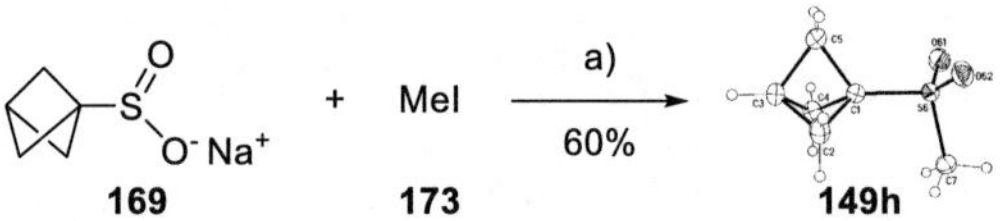

Scheme 53. Alkylation of BCP sulfinate **169** with iodomethane (**173**). The structure of **149h** could be determined by single-crystal X-ray diffraction. Displacement parameters are drawn at 50% probability level. Crystallographic data can be found in chapter 7.4. *a)* 1.00 equiv. **169**, 1.50 equiv. **173**, DMF, 0 °C, 4 h.

An oxidation of **169** with hydrogen peroxide and a tungsten catalyst was not successful (Scheme 54).

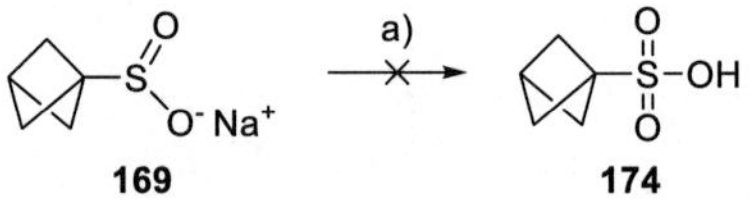

Scheme 54. Unsuccessful attempt to oxidize sulfinate **169** to the corresponding sulfonic acid **174**. *a)* 1.00 equiv. **169**, 3.00 equiv. H$_2$O$_2$, 20 mol% Na$_2$WO$_4$, H$_2$O, rt, 20 h.

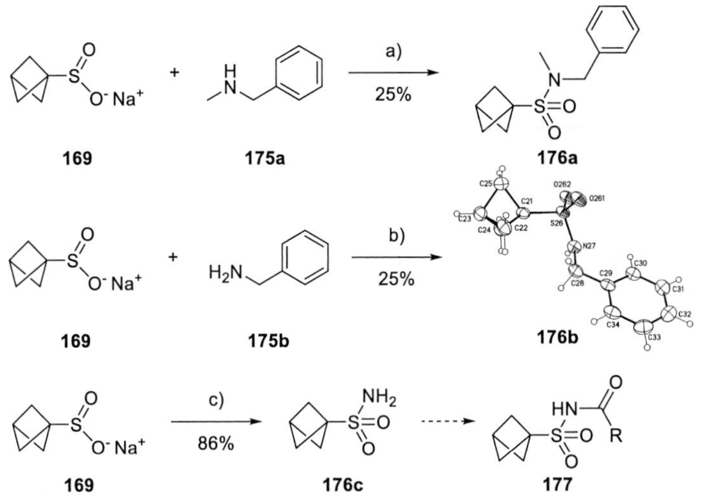

Scheme 55. Synthesis of BCP sulfonamides **176** from the sulfinate **169**. The structure of **176b** could be determined by single-crystal X-ray diffraction. Displacement parameters are drawn at 50% probability level. Crystallographic data can be found in chapter 7.4. *a)* 1.00 equiv. **169**, 1.30 equiv. **175a**, 1.00 equiv. I_2, H_2O, rt, 48 h; *b)* 1.00 equiv. **169**, 1.50 equiv. **175b**, 1.00 equiv. I_2, H_2O, rt, 48 h; *c)* 1.00 equiv. **169**, 2.00 equiv. NH_2OSO_3H, 1.00 equiv. KOAc, H_2O, rt, 18 h.

Returning to the initial motivation to synthesize the BCP sulfinate **169**, the application in the sulfonamide synthesis was tested. Therefore, simple and mild conditions by Yuan *et al.* were chosen.[148] This reaction is based on the iodination of the sulfinate to generate the highly reactive sulfonyl iodide. A secondary and a primary amine could be converted to the BCP sulfonamides **176a** and **176b** successfully (Scheme 55). Although the yield of 25% for both cases is low, compared to the literature, it's still a proof-of-principle and the first synthesis of such structures. The structure of **176b** could be proven by single-crystal X-ray diffraction. The primary sulfonamide **176c** could be obtained in a very good yield of 86% with the aminating reagent hydroxylamine-*O*-sulfonic acid. This product is an interesting building block that could be applied in the synthesis of *N*-acyl sulfonamides **177**, another interesting motif in medicinal chemistry.[149]

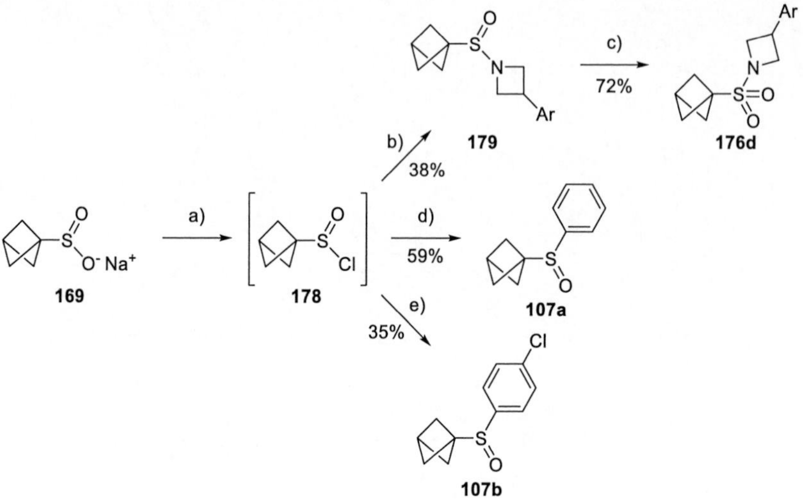

Scheme 56. Chlorination of **169** and reactions of the obtained sulfinyl chloride **178**. *a)* 1.00 equiv. **169**, 1.50 equiv. SOCl₂, 10 mol% DMF, CH₂Cl₂, rt, 16 h; *b)* 2.00 equiv. crude **178** in CH₂Cl₂, 1.00 equiv. R₂NH, 2.00 equiv. Et₃N, CH₂Cl₂, rt, 5 d; *c)* 1.00 equiv. **179**, 1.50 equiv. *m*CPBA, CH₂Cl₂, rt, 30 min; *d)* 1.00 equiv. crude **178** in CH₂Cl₂, 2.50 equiv. PhMgBr, 0 °C to rt, 1 h; *e)* 1.00 equiv. crude **178** in CH₂Cl₂, 2.01 equiv. 4-chlorophenylmagnesium bromide, 0 °C to rt, 1 h.

Through chlorination the sulfinate **169** could be converted into sulfinyl chloride **178**, which was directly used in the synthesis of sulfinamide **179** (Scheme 56). The sulfinamide could be oxidized to the corresponding sulfonamide **176d** using *m*CPBA. Both structures can unfortunately not be shown completely due to confidentiality reasons.[7] The reaction with Grignard reagents was successful as well and opens another variety of nucleophiles that can react with **178**. Sulfoxide **107b** was employed in the synthesis of BCP sulfoximine **108b** in chapter 5.3.2.

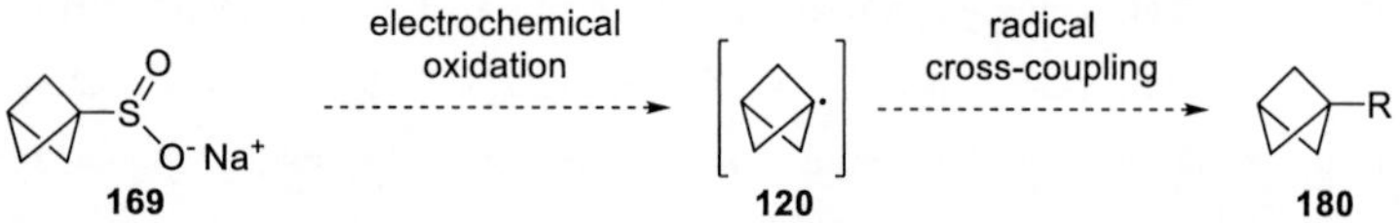

Scheme 57. Ongoing investigations in collaboration with Dr. Kevin Lam (University of Greenwich, UK) towards electrochemical oxidation of the sulfinate **169** to obtain radical intermediate **120**.

An application of sulfinate **169** in desulfinylative reactions would be highly desirable (Scheme 57). There are methods available for such a transformation, but none of them work for simple tertiary alkyl sulfinates.[150-151] All of the successful tertiary examples are cyclopropanes or α-fluorinated

---

[7] The compounds **179** and **176d** were synthesized and characterized during a stay at Boehringer Ingelheim Pharma GmbH & Co. KG.

compounds. Baran *et al.* synthesized a cubane sulfinate but did not show any further reactions with it.[152] Knauber *et al.* mentioned in their publication that *tert*-butanesulfinate did not afford the desired product.[153] In a collaborative project with Dr. Kevin Lam (University of Greenwich) we aim at a novel electrochemical procedure to perform desulfinylative cross-couplings with sulfinate **169**. A possible extension to 3-substituted BCP sulfinates that may form more stable intermediates is planned (see chapter 6).

One of the aims of this thesis was the development of a bench-stable building block for the synthesis of sulfur BCPs. The sulfinate **169** presented in this chapter offers a variety of applications, mainly in the synthesis of BCP sulfones and sulfonamides. It was available in a multigram scale synthesis from **1** without the need for sophisticated purification methods. Further modifications and 3-substituted BCP sulfinates will be part of subsequent studies.

6 Summary and outlook

Bicyclo[1.1.1]pentanes (BCPs) are one of the non-classical bioisosteres that recently gained a lot of interest in drug discovery as non-conjugated rigid hydrocarbons (NRHs). There has been a substantial progress in syntheses of BCPs through CC and CN bond formation with the precursor [1.1.1]propellane (**1**). The CS bond formation with **1** has been underdeveloped and therefore not used in medicinal chemistry. In this work methods were investigated and developed for the synthesis of BCP sulfides **104–106** from **1** using thiols or disulfides (Scheme 58).

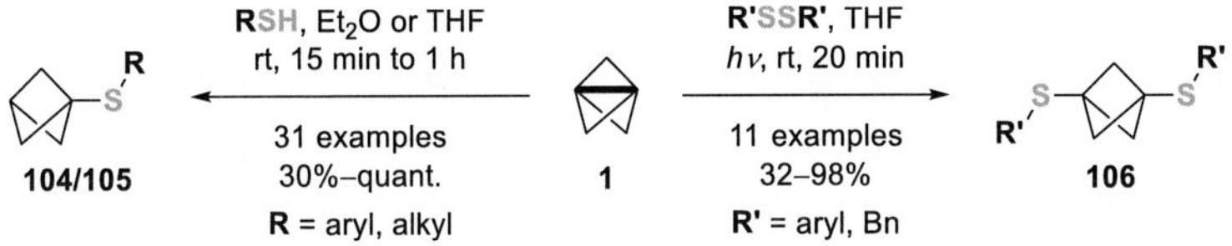

RSH, Et$_2$O or THF		R'SSR', THF
rt, 15 min to 1 h		hv, rt, 20 min
31 examples	**1**	11 examples
30%–quant.		32–98%
R = aryl, alkyl		R' = aryl, Bn

104/105   **106**

Scheme 58. Summary of the developed thiol addition to **1** and the insertion of **1** into disulfide bonds.

Both reactions proceed through a radical mechanism and show good functional group tolerance. The only limitation that was discovered for the thiol addition to **1** are heteroaromatic thiols. The insertion of **1** into disulfides as shown in this thesis is mainly limited to aromatic disulfides, but could possibly be extended towards alkyl disulfides by employing a radical initiator. If a mixture of disulfides is used in the reaction, a product mixture can be obtained with the non-symmetrically substituted BCP as the major product.

The obtained products **104–106** were modified to obtain more drug-like compounds. Oxidation to sulfoxides **107** and subsequent imination to sulfoximines **108** were shown in proof-of-concept reactions. An enantioselective oxidation would lead to enantiopure sulfoximines and should be part of a future project. A one step procedure from the sulfides to the corresponding sulfoximines **108** was applied to a variety of compounds (Scheme 59), together with the master student Lukas Langer (Bräse group). A modification of these products **108** at the nitrogen atom was also investigated. A literature procedure for the *N*-arylation of sulfoximines was optimized and applied in a broad substrate scope. Further, BCP sulfone was found to be a directing group for *ortho*-lithiation similar to *tert*-butyl sulfone.

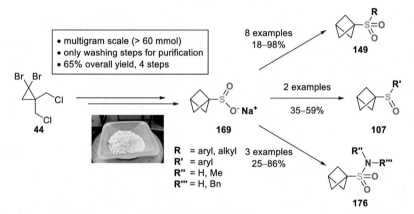

Scheme 59. Summary of the sulfoximine synthesis and subsequent *N*-arylation to **159**. This part was done together with the master student Lukas Langer.

The final aim of this thesis was the synthesis and application of a bench-stable precursor for sulfur BCPs. Therefore, BCP sulfinate **169** was synthesized in a four step procedure from commercially available cyclopropane **44** (Scheme 60). This synthetic route delivered **169** in 65% overall yield on a multigram scale without any chromatographic purification or crystallization. The application of the building block was shown in a broad range of reactions, leading to BCP sulfones **149**, sulfoxides **107**, a sulfinamide **179** and sulfonamides **176**.

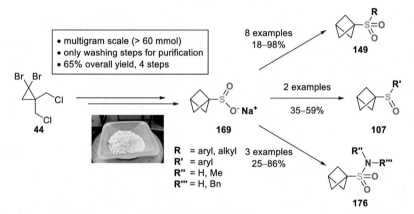

Scheme 60. Synthesis of BCP sulfinate **169** on a multigram scale and application of **169** in the synthesis of BCP sulfones **149**, sulfoxides **107** and sulfonamides **176**.

In a collaborative project with Dr. Kevin Lam (University of Greenwich, UK) the sulfinate **169** is currently under investigation as precursor for a desulfinylative radical-radical cross-coupling reaction induced by electrochemical oxidation. As the sulfinate **169** was synthesized during a collaboration with Boehringer Ingelheim Pharma GmbH & Co. KG the building block is now part of the company compound library and will hopefully find application in many drug discovery projects. An extension to 3-substituted BCP sulfinates is the next step and would enable a much broader application in bioisosteric replacements. The opening of **1** with Grignard reagents or organolithium compounds and subsequent reaction of the metallated BCP **182** with SO_2 (or a SO_2 surrogate like DABSO[154]) should enable access to such building blocks (Scheme 61).

The sulfinate **169** still offers interesting modifications that could be explored in future projects. Willis *et al.* showed that the formation of trimethylsilyl sulfinates like **184** changes the selectivity in the reaction with nucleophiles towards the sulfur (Scheme 61).[155] Therefore, the synthesis of BCP sulfoxides **107** without chlorination of the starting material could be possible. This method might also be applicable in an enantioselective manner.

During the thiol addition to **1** a tetrazole derivative was planned, which could be a useful building block for desulfinylative cross-coupling reactions. The synthesis was not successful, but the tetrazole sulfone product **119** might be accessible starting from the sulfinate **169** (Scheme 61).

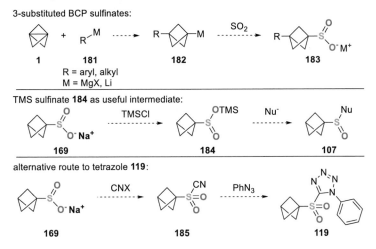

Scheme 61. Possible future projects around BCP sulfinates.

The use of BCPs as bioisosteres significantly increased during the last decade. The rising interest led to the development of many useful synthetic methods. However, easy access to 2-substituted [1.1.1]propellanes is still a major goal in this field. This would not only enable more sophisticated bioisosteric replacements, but also the use of [1.1.1]propellanes for bioconjugations, e.g. with disulfides. The basis of such a conjugative reaction could be the insertion of **1** into disulfides as described in this thesis.

7   Experimental part

7.1   General remarks

7.1.1   Preparative work

The starting materials, solvents and reagents were purchased from ABCR, Acros, Alfa Aesar, Carbolution, ChemPUR, Fluka, Iris, Merck, Riedel-de Haën, TCI, Thermo Fisher Scientific, Sigma Aldrich and used without further purification. Thiols were checked by GC-MS before use and purified if necessary. For this purpose, the thiol was dissolved in diethyl ether, deprotonated by washing with 1 M NaOH solution and then extracted from the aqueous phase that has been acidified again with conc. HCl solution. After drying over magnesium sulfate the solvent was removed under reduced pressure and the pure thiol was obtained. Volatile thiols were distilled before use. Fmoc-cysteamine (**132h**) was provided by Anne Schneider (Bräse group).

All reactions containing air- and moisture-sensitive compounds were performed under argon atmosphere using oven-dried glassware applying Schlenk techniques. Liquids were added *via* steel cannulas and solids were added directly in powdered shape.

Reactions were accomplished at room temperature, if nothing else is mentioned. For low reaction temperatures flat dewars with ice/water or isopropanol/dry ice mixtures were used.

Solvents were removed at 40 °C with a rotary evaporator under reduced pressure. For solvent mixtures each solvent was measured volumetrically. If nothing else is mentioned, saturated, aqueous solutions of inorganic salts were used.

Crude products were purified *via* flash chromatography using Merck silica gel 60 (0.040 × 0.063 mm, 230–400 mesh ASTM) and quartz sand (glowed and purified with hydrochloric acid). Therefore, the eluents were distilled or used directly in *p.a.* quality bought from Merck.

7.1.2   Solvents and reagents

Solvents of technical quality were purified by distillation or with the solvent purification system MB SPS5 from MBRAUN prior to use. Solvents of the grade *p.a.* were purchased (Acros, Fisher Scientific, Sigma Aldrich, Roth, Riedel-de Haën) and were used without further purification. Absolute solvents were dried, using the methods listed in Table 10 and were stored under argon afterwards or were purchased from a commercial supplier (abs. acetonitrile (Acros, <0.005% water), anhydrous *N,N*-dimethylformamide (Sigma Aldrich, <0.005% water), anhydrous dimethyl sulfoxide (Sigma Aldrich, <0.005% water), abs. methanol (Fischer, <0.005% water)).

Table 10. Methods for the absolutizing of solvents. All distillations were carried out under argon atmosphere.

Solvent	Method
Dichloromethane	heating to reflux over calcium hydride, distilled over a packed column or MB SPS5
Tetrahydrofuran	heating to reflux over sodium metal (benzophenone as an indicator), distilled over a packed column or MB SPS5
Diethyl ether	heating to reflux over sodium metal (benzophenone as an indicator) distilled over a packed column or MB SPS5

7.1.3 Analytics and equipment

7.1.3.1 Nuclear magnetic resonance (NMR)

All NMR spectra were recorded, using the following machines:

^{1}H NMR: BRUKER *Avance AV 300* (300 MHz), BRUKER *Avance 400* (400 MHz), BRUKER *Avance DRX 500* (500 MHz). The chemical shift δ is expressed in parts per million (ppm) where the residual signal of the solvent was used as reference: chloroform-d_1 ($\delta = 7.26$ ppm), dimethyl sulfoxide-d_6 ($\delta = 2.50$ ppm) or deuterium oxide-d_2 ($\delta = 4.79$ ppm).[156] The spectra were analyzed according to first order.

^{13}C NMR: BRUKER *Avance 300* (75 MHz), BRUKER *Avance 400* (100 MHz), BRUKER *Avance DRX 500* (125 MHz). The chemical shift δ is expressed in parts per million (ppm) where the residual signal of the solvent was used as reference: chloroform-d_1 ($\delta = 77.0$ ppm) or dimethyl sulfoxide-d_6 ($\delta = 39.4$ ppm).[156] The spectra were ^{1}H-decoupled and characterization of the ^{13}C NMR-spectra ensured through the DEPT-technique (DEPT = Distortionless Enhancement by Polarization Transfer) and are stated as follows: DEPT: "+" = primary or secondary carbon atoms (positive DEPT-signal), "–" = secondary carbon atoms (negative DEPT-signal), C_q = quaternary carbon atoms (no DEPT-signal).

All spectra were obtained at room temperature. NMR-solvents were obtained from Eurisotop and Sigma Aldrich: chloroform-d_1, dimethylsulfoxide-d_6, deuterium oxide-d_2. For central symmetrical signals the midpoint and for multiplets the range of the signal region are given. The multiplicities of the signals are abbreviated as follows: s = singlet, d = doublet, t = triplet, q = quartet, hept = heptet, bs = broad singlet, m = multiplet, b = broad and combinations thereof. All coupling constants 'J' are stated as modulus in Hertz [Hz].

In some cases the signals were assigned using ^{1}H-^{13}C-HSQC (Heteronuclear Single Quantum Coherence) and ^{1}H-^{13}C-HMBC (Heteronuclear Multiple Quantum Correlation) techniques.

7.1.3.2 Infrared spectroscopy (IR)

IR-spectra were recorded on a BRUKER *Alpha P* and a BRUKER *IFS 88*. Measurements of all samples were conducted *via* attenuated total reflection (ATR). The position of the absorption bands is given as wavenumber $\tilde{v}$ with the unit $[cm^{-1}]$.

7.1.3.3 Mass spectrometry (GC-MS, EI-MS, FAB-MS, ESI-MS, HRMS)

GC-MS (Gas chromatography-mass spectrometry). The measurements were recorded with an Agilent Technologies model 6890N (electron impact ionization), equipped with a Agilent 19091S-433 column (5% phenyl methyl siloxane, 30 m, 0.25 μm) and a 5975B VL MSD detector with a turbo pump. Helium was used as a carrier gas.

EI-MS and **FAB-MS**. The measurements were recorded with a Finnigan MAT 95 (70 eV). Ionization was achieved through either EI (electron ionization) or FAB (fast atom bombardment).

ESI-MS. The measurements were recorded with a ThermoFisher QExactive Plus (4 kV) with a ThermoFisher LT Orbitrap XL. Ionization was achieved through ESI (electrospray ionization).

HR-MS (high resolution-mass spectra). The measurements were either recorded with the Finnigan MAT 95 (EI/FAB) or with the ThermoFisher QExactive Plus (ESI). The following abbreviations were used: calc. = expected value (calculated); found = value found in analysis.

Notation of molecular fragments is given as mass to charge ratio (m/z); the intensities of the signals are noted in percent relative to the base signal (100%). As abbreviation for the ionized molecule $[M]^+$ is used. Characteristic fragmentation peaks are given as $[M-fragment]^+$ and $[fragment]^+$.

7.1.3.4 Single-crystal X-ray diffraction (XRD)/ Powder X-ray diffraction (PXRD)

Two different diffractometers were used in this work.

Bruker D8 Venture diffractometer with Photon100 detector at 123(2) K using Cu-Kα radiation ($\lambda = 1.54178$ Å). Dual space methods (SHELXT)[157] were used for structure solution and refinement was carried out using SHELXL-2014 (full-matrix least-squares on F^2).[158] Hydrogen atoms were localized by difference electron density determination and refined using a riding model (H(O) free). Semi-empirical absorption corrections were applied.

STOE STADI VARI diffractometer at 200 K with monochromated Ga-Kα radiation ($\lambda = 1.34143$ Å). Using Olex2,[159] the structure was solved with the ShelXS[160] structure solution program using direct methods and refined with the ShelXL[158] refinement package using least squares minimization. Refinement was performed with anisotropic temperature factors for all non-hydrogen atoms; hydrogen atoms were calculated on idealized positions.

7.1.3.5 Analytical high-performance liquid chromatography (HPLC)

The determination of the purity of the compounds was carried out on an Agilent 1100 series HPLC system with a G1322A degasser, a G1311A pump, G1313A autosampler, a G1316A column oven and a G1315B diode array detector. The flow rate was 1 mL/min. The stationary phase used was a VDSpher C18-M-SE (VDS Optilab) C18 column (5 µm, 4.0 mm × 250 mm). The runs were carried out with a linear gradient of A: 5% acetonitrile, 0.1% TFA in water to B: 95% acetonitrile, 0.1% TFA in water within 30 min. The purity was determined by integration of the signals at 218 nm or 256 nm.

7.1.3.6 Preparative HPLC

The purification of some compounds was carried out on a Shimadzu SPD-M10AVP series HPLC system with LC-8A pumps, a SCL-10AVP diode array detector and a type 202 fraction collector from Gilson. The flow rate was 15 mL / min. The stationary phase used was a VDSphereR C18-E (VDS Optilab) C18 column (5 µm, 20 mm × 250 mm). The runs were adjusted to the corresponding compounds. All separation were performed with the following eluents: A = ddH$_2$O, 0.1% TFA; B = acetonitrile, 0.1% TFA. The separation of the crude products was detected at 230 nm, 256 nm, 280 nm, 300 nm and 400 nm.

7.1.3.7 Melting point determination

Melting points were detected on an OptiMelt MPA100 device from the company Stanford Research System.

7.1.3.8 Thin layer chromatography (TLC)

All reactions were monitored by TLC using silica gel coated aluminium plates (Merck, silica gel 60, F$_{254}$). The detection was performed with UV light (254 nm) and/or dipped into a solution of Seebach reagent (2.5% phosphor molybdic acid, 1.0% Cerium(IV) sulfate tetrahydrate and 6.0% sulfuric acid in H$_2$O, dipping solution) or potassium permanganate (1.5 g KMnO$_4$, 10 g K$_2$CO$_3$ and 1.25 mL 10% NaOH in 200 mL H$_2$O, dipping solution) and heated with a heat gun.

7.1.3.9 Analytical scales

For mass determination a balance from Satorius (LC 620 S) was used.

7.2   Syntheses and characterizations

7.2.1   General procedures (GPs)

GP1: Thiol addition to [1.1.1]propellane

A solution of [1.1.1]propellane (**1**) (1.00 equiv. in Et$_2$O or Et$_2$O/*n*-pentane) was added to a solution of the thiol **110/132** (1.00–1.25 equiv.) in Et$_2$O or THF under argon atmosphere at room temperature and stirred for 15 min to 1 h. The reaction mixture was diluted with *n*-pentane, washed with 1 M NaOH solution, dried by the addition of Na$_2$SO$_4$, filtered and evaporated. If necessary, the product was purified *via* column chromatography.

GP2: Insertion of [1.1.1]propellane into disulfide bonds (BCPs)

A solution of [1.1.1]propellane (**1**) (1.00 equiv. in Et$_2$O) was added to a solution of the disulfide **138** (3.00 equiv.) in THF under argon atmosphere in a quartz flask at room temperature and the mixture was irradiated with UV light (256 nm, 500 W) for 20 min. After evaporation of the solvent, the crude residue was purified *via* column chromatography.

GP3: Insertion of [1.1.1]propellane into disulfide bonds ([2]staffanes)

A solution of [1.1.1]propellane (**1**) (2.00 equiv. in Et$_2$O) was added to a solution of the disulfide **138** (1.00 equiv.) in THF under argon atmosphere in a quartz flask at room temperature and the mixture was irradiated with UV light (256 nm, 500 W) for 20 min. After evaporation of the solvent, the crude residue was purified *via* column chromatography or preparative TLC.

GP4: Synthesis of BCP sulfoximines

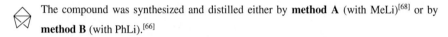

104/105   **108**

The sulfide **104/105** (1.00 equiv.) was dissolved in MeOH and stirred at room temperature (open flask). First, ammonium carbonate (2.00 equiv.) was added, followed by (diacetoxyiodo)benzene (3.00 equiv.). The reaction mixture was stirred for 30 min. The solvent was removed under reduced pressure and the crude product was purified *via* column chromatography.

7.2.2   Thiol addition to [1.1.1]propellane

[1.1.1]Propellane (**1**)

The compound was synthesized and distilled either by **method A** (with MeLi)[68] or by **method B** (with PhLi).[66]

Method A: In a flame-dried round-bottomed flask that has been purged with argon 1,1-dibromo-2,2-bis(chloromethyl)cyclopropane (**44**, 90% purity, 2.22 g, 6.74 mmol, 1.00 equiv.) was dissolved in *n*-pentane (2.50 mL) and cooled to –78 °C. A 1.6 M MeLi solution (8.50 mL, 299 mg, 13.6 mmol, 2.02 equiv.) in Et$_2$O was added dropwise under vigorous stirring. After complete addition the reaction mixture was maintained at –78 °C for 15 min before being allowed to warm to 0 °C and stirred at this temperature for further 1.5 h. A short distillation bridge that has been evacuated and purged with argon three times, was attached to the three necked flask. The receiver flask was cooled to –78 °C and the volatiles were transferred under reduced pressure. A solution of **1** and MeBr in Et$_2$O/*n*-pentane was obtained and stored at –78 °C under argon atmosphere.

Method B: In a flame-dried round-bottomed flask that has been purged with argon 1,1-dibromo-2,2-bis(chloromethyl)cyclopropane (**44**, 90% purity, 2.22 g, 6.74 mmol, 1.00 equiv.) was dissolved in diethyl ether (8.50 mL) and cooled to –40 °C. A 1.9 M PhLi solution (7.16 mL, 1.14 g, 13.6 mmol, 2.02 equiv.) in Bu$_2$O was added dropwise under vigorous stirring. After complete addition the mixture was allowed to warm to 0 °C and stirred at this temperature for further 2 h. The reaction flask was attached to an argon purged rotavap with dry ice condenser. The receiving flask was cooled to –78 °C and the product was distilled together with diethyl ether. The water bath was set to 20 °C and the pressure was reduced from 500 mbar to 20 mbar slowly. A solution of **1** in Et$_2$O was obtained and stored at –78 °C under argon atmosphere.

To quantify the concentration of **1** an aliquot was taken and reacted with thiophenol (see **104a**). With the calculated concentration the yield could be determined retrospectively and ranged from 21% (0.13 M, method A) to 76% (0.60 M, method B) in different batches.

^{1}H NMR (300 MHz, CDCl$_3$): δ = 2.00 (s, 6H, 3 x CH_2) ppm; ^{13}C NMR (75 MHz, CDCl$_3$): δ = 74.1 (–, 3 x CCH_2), 1.0 (C$_q$, 2 x CCH_2) ppm.

The analytical data is in accordance with the literature.[57]

1-(Phenylthio)-bicyclo[1.1.1]pentane (**104a**)

In an argon flushed 10 mL flask a 1.0 M solution of thiophenol (**110a**) in Et$_2$O (0.68 mL, 680 µmol) was added to 1.00 mL of the stock solution of **1** (prepared by method A or B) with unknown concentration. The reaction was stirred for 15 min at room temperature. The mixture was diluted with 10 mL n-pentane and washed with 10 mL of a 1 M NaOH solution. The organic phase was dried over Na$_2$SO$_4$, filtrated and the solvent was removed under reduced pressure to obtain the product as a pale yellow oil. The turnover of this reaction is assumed to be quantitative to calculate the concentration of the solution of **1**.[68, 73]

^{1}H NMR (400 MHz, CDCl$_3$): δ = 7.46–7.43 (m, 2H, Ar-H), 7.33–7.26 (m, 3H, Ar-H), 2.73 (s, 1H, CH), 1.96 (s, 6H, 3 × CH_2) ppm; ^{13}C NMR (100 MHz, CDCl$_3$): δ = 134.2 (C$_q$, C_{Ar}), 133.6 (+, 2 × CH_{Ar}), 128.9 (+, 2 × CH_{Ar}), 127.6 (+, CH_{Ar}), 54.1 (–, 3 × CH_2), 45.8 (C$_q$, C$_{Ar}$SC), 28.8 (+, CH) ppm; IR (ATR): $\tilde{v}$ = 2998, 2909, 2874, 1583, 1472, 1438, 1203, 1128, 1088, 1066, 1024, 894, 777, 741, 691, 548, 502, 423, 385 cm^{-1}; MS (EI, 50 °C): m/z (%) = 176 (18) [M]$^+$, 135 (42) [M–C$_3$H$_5$]$^+$, 109 (71) [M–C$_5$H$_7$]$^+$, 99 (11) [M–C$_6$H$_5$]$^+$, 78 (64) [C$_6$H$_5$+H]$^+$, 77 (39) [C$_6$H$_5$]$^+$, 67 (97) [C$_5$H$_7$]$^+$, 41 (100) [C$_3$H$_5$]$^+$; HRMS (EI, 50 °C): calc. for C$_{11}$H$_{12}{}^{32}$S [M]$^+$ 176.0654; found 176.0655.

The analytical data is in accordance with the literature.[73]

1-(4-Bromophenylthio)-bicyclo[1.1.1]pentane (**104c**)

According to **GP1** 4-bromothiophenol (**110c**, 51.0 mg, 270 µmol, 1.00 equiv.) was reacted with **1** (1.00 mL, 270 µmol, 1.00 equiv.) in Et$_2$O (270 µL) for 15 min. The product was obtained as a yellow oil (45.0 mg, 176 µmol, 65%).

^{1}H NMR (500 MHz, CDCl$_3$): δ = 7.44–7.41 (m, 2H, Ar-H), 7.31–7.28 (m, 2H, Ar-H), 2.73 (s, 1H, CH), 1.95 (s, 6H, 3 × CH_2) ppm; ^{13}C NMR (125 MHz, CDCl$_3$): δ = 135.2 (+, 2 × CH_{Ar}), 133.4 (C$_q$, C$_{Ar}$S), 132.0 (+, 2 × CH_{Ar}), 121.9 (C$_q$, C$_{Ar}$Br), 54.1 (–, 3 × CH_2), 45.6 (C$_q$, C$_{Ar}$SC), 28.9 (+, CH) ppm; IR (ATR): $\tilde{v}$ = 2979, 2910, 2875, 1561, 1471, 1384, 1205, 1128, 1090, 1069, 1009, 895,

815, 776, 729, 549, 511, 492, 445 cm^{-1}; MS (EI, 20 °C): m/z (%) = 256/254 (26/26) [M]$^+$, 190/188 (30/29) [M–C$_5$H$_7$+H]$^+$, 134 (54) [M–C$_3$H$_5$–Br]$^+$, 109 (32) [M–C$_5$H$_7$–Br+H]$^+$, 108 (40) [M–C$_5$H$_7$–Br]$^+$, 67 (100) [C$_5$H$_7$]$^+$; HRMS (EI, 20 °C): calc. for C$_{11}$H$_{11}$^{79}Br^{32}S [M]$^+$ 253.9759; found 253.9759.

1-(4-Methylphenylthio)-bicyclo[1.1.1]pentane (104d)

According to **GP1** 4-methylthiophenol (**110d**, 33.5 mg, 270 μmol, 1.00 equiv.) was reacted with **1** (1.00 mL, 270 μmol, 1.00 equiv.) in Et$_2$O (270 μL) for 15 min. The product was obtained as a yellow oil (45.0 mg, 236 μmol, 87%).

^{1}H NMR (500 MHz, CDCl$_3$): δ = 7.34–7.32 (m, 2H, Ar-H), 7.12–7.10 (m, 2H, Ar-H), 2.71 (s, 1H, CH), 2.34 (s, 3H, CH_3), 1.92 (s, 6H, 3 × CH_2) ppm; ^{13}C NMR (125 MHz, CDCl$_3$): δ = 137.7 (C$_q$, C_{Ar}CH$_3$), 134.0 (+, 2 × CH$_{Ar}$), 130.4 (C$_q$, C_{Ar}S), 129.7 (+, 2 × CH$_{Ar}$), 54.0 (–, 3 × CH$_2$), 45.9 (C$_q$, C$_{Ar}$SC), 28.7 (+, CH), 21.3 (+, CH$_3$) ppm; IR (ATR): $\tilde{v}$ = 2978, 2910, 2874, 1490, 1447, 1398, 1205, 1129, 1093, 1018, 890, 808, 775, 733, 706, 549, 507, 449 cm^{-1}; MS (EI, 20 °C): m/z (%) = 191 (13) [M+H]$^+$, 190 (87) [M]$^+$, 175 (10) [M–CH$_3$]$^+$, 149 (100) [M–C$_3$H$_5$]$^+$, 134 (32) [M–C$_3$H$_5$–CH$_3$]$^+$, 124 (85) [M–C$_5$H$_7$+H]$^+$, 123 (30) [M–C$_5$H$_7$]$^+$, 91 (87) [M–C$_5$H$_7$S]$^+$, 67 (52) [C$_5$H$_7$]$^+$; HRMS (EI, 20 °C): calc. for C$_{12}$H$_{14}$^{32}S [M]$^+$ 190.0811; found 190.0812.

1-(4-*tert*-Butylphenylthio)-bicyclo[1.1.1]pentane (104e)

According to **GP1** 4-*tert*-butylthiophenol (**110e**, 44.9 mg, 270 μmol, 1.00 equiv.) was reacted with **1** (1.00 mL, 270 μmol, 1.00 equiv.) in Et$_2$O (270 μL) for 15 min. The product was obtained as a pale yellow oil (50.0 mg, 215 μmol, 79%).

^{1}H NMR (400 MHz, CDCl$_3$): δ = 7.38–7.35 (m, 2H, Ar-H), 7.33–7.30 (m, 2H, Ar-H), 2.71 (s, 1H, CH), 1.95 (s, 6H, 3 × CH_2), 1.31 (s, 9H, 3 × CH_3) ppm; ^{13}C NMR (100 MHz, CDCl$_3$): δ = 150.7 (C$_q$, C_{Ar}CCH$_3$), 133.4 (+, 2 × CH$_{Ar}$), 130.6 (C$_q$, C_{Ar}S), 125.9 (+, 2 × CH$_{Ar}$), 54.1 (–, 3 × CH$_2$), 45.8 (C$_q$, C$_{Ar}$SC), 34.7 (C$_q$, CCH$_3$), 31.4 (+, 3 × CH$_3$), 28.8 (+, CH) ppm; IR (ATR): $\tilde{v}$ = 2961, 2907, 2873, 1488, 1460, 1393, 1362, 1266, 1203, 1129, 1116, 1013, 895, 827, 742, 561, 410 cm^{-1}; MS (EI, 20 °C): m/z (%) = 232 (42) [M]$^+$, 217 (100) [M–CH$_3$]$^+$, 151 (25) [M–CH$_3$–C$_5$H$_7$+H]$^+$, 67 (11) [C$_5$H$_7$]$^+$, 57 (27) [C(CH$_3$)$_3$]$^+$; HRMS (EI, 20 °C): calc. for C$_{15}$H$_{20}$^{32}S [M]$^+$ 232.1280; found 232.1279.

1-(4-Methoxyphenylthio)-bicyclo[1.1.1]pentane (**104f**)

According to **GP1** 4-methoxythiophenol (**110f**, 84.0 mg, 599 μmol, 1.01 equiv.) was reacted with **1** (1.00 mL, 596 μmol, 1.00 equiv.) in Et$_2$O (600 μL) for 15 min. The product was obtained as a pale yellow oil (111 mg, 538 μmol, 90%).

^{1}H NMR (400 MHz, CDCl$_3$): δ = 7.40–7.35 (m, 2H, Ar-H), 6.86–6.82 (m, 2H, Ar-H), 3.81 (s, 3H, CH_3), 2.69 (s, 1H, CH), 1.88 (s, 6H, 3 × CH_2) ppm; ^{13}C NMR (100 MHz, CDCl$_3$) δ = 159.7 (C$_q$, C$_{Ar}$OCH$_3$), 136.0 (+, 2 × CH$_{Ar}$), 124.5 (C$_q$, C$_{Ar}$S), 114.5 (+, 2 × CH$_{Ar}$), 55.4 (+, CH$_3$), 53.8 (–, 3 × CH$_2$), 46.3 (C$_q$, C$_{Ar}$SC), 28.5 (+, CH) ppm; IR (ATR): $\tilde{v}$ = 2977, 2908, 2874, 2835, 1590, 1571, 1491, 1462, 1440, 1284, 1242, 1205, 1171, 1130, 1096, 1030, 892, 826, 798, 640, 628, 550, 525, 457 cm^{-1}; MS (EI, 20 °C): m/z (%) = 206 (100) [M]$^+$, 165 (23) [M–C$_3$H$_5$]$^+$, 140 (66) [M–C$_5$H$_7$+H]$^+$, 139 (35) [M–C$_5$H$_7$]$^+$, 125 (33) [M–C$_5$H$_7$–CH$_3$+H]$^+$, 124 (9) [M–C$_5$H$_7$–CH$_3$]$^+$, 121 (45) [M–C$_4$H$_7$–OCH$_3$]$^+$, 67 (18) [C$_5$H$_7$]$^+$; HRMS (EI, 20 °C): calc. for C$_{12}$H$_{14}$O^{32}S [M]$^+$ 206.0765; found 206.0764.

1-(4-Nitrophenylthio)-bicyclo[1.1.1]pentane (**104g**)

According to **GP1** 4-nitrothiophenol (**110g**, 93.0 mg, 599 μmol, 1.01 equiv.) was reacted with **1** (1.00 mL, 596 μmol, 1.00 equiv.) in THF (1.20 mL) for 15 min. The product was obtained as a yellow solid (62 mg, 280 μmol, 47%).

^{1}H NMR (400 MHz, CDCl$_3$): δ = 8.15–8.12 (m, 2H, Ar-H), 7.53–7.49 (m, 2H, Ar-H), 2.84 (s, 1H, CH), 2.13 (s, 6H, 3 × CH_2) ppm; ^{13}C NMR (100 MHz, CDCl$_3$): δ = 146.2 (C$_q$, C$_{Ar}$NO$_2$), 145.6 (C$_q$, C$_{Ar}$S), 130.7 (+, 2 × CH$_{Ar}$), 123.9 (+, 2 × CH$_{Ar}$), 54.7 (–, 3 × CH$_2$), 44.6 (C$_q$, C$_{Ar}$SC), 30.0 (+, CH) ppm; IR (ATR): $\tilde{v}$ = 2983, 2913, 2877, 1594, 1575, 1475, 1336, 1259, 1208, 1124, 1090, 1012, 892, 851, 742, 684, 535, 408 cm^{-1}; MS (EI, 30 °C): m/z (%) = 222 (6) [M+H]$^+$, 221 (38) [M]$^+$, 180 (7) [M–C$_3$H$_5$]$^+$, 134 (31) [M–C$_3$H$_5$–NO$_2$]$^+$, 67 (100) [C$_5$H$_7$]$^+$; HRMS (EI, 30 °C) calc. for C$_{11}$H$_{11}$O$_2$N^{32}S [M]$^+$ 221.0505; found 221.0506.

1-(4-Aminophenylthio)-bicyclo[1.1.1]pentane (**104h**)

According to **GP1** 4-aminothiophenol (**110h**, 129 mg, 1.03 mmol, 1.25 equiv.) was reacted with **1** (1.40 mL, 826 μmol, 1.00 equiv.) in Et$_2$O (1.00 mL) for 15 min. The product was obtained as a yellow oil (86 mg, 450 μmol, 54%).

^{1}H NMR (400 MHz, CDCl$_3$): δ = 7.26–7.23 (m, 2H, Ar-H), 6.63–6.61 (m, 2H, Ar-H), 3.70 (b, 2H, NH$_2$), 2.68 (s, 1H, CH), 1.86 (s, 6H, 3 × CH$_2$) ppm; ^{13}C NMR (100 MHz, CDCl$_3$): δ = 146.5 (C$_q$, C_{Ar}NH$_2$), 136.2 (+, 2 × CH$_{Ar}$), 121.5 (C$_q$, C_{Ar}S), 115.5 (+, 2 × CH$_{Ar}$), 53.7 (–, 3 × CH$_2$), 46.4 (C$_q$, C_{Ar}SC), 28.4 (+, CH) ppm; IR (ATR): $\tilde{v}$ = 3431, 3342, 3213, 3021, 2971, 2905, 2872, 1884, 1631, 1593, 1491, 1423, 1296, 1201, 1177, 1125, 1096, 1063, 1006, 935, 890, 814, 771, 646, 546, 517, 423, 406 cm^{-1}; MS (EI, 40 °C): m/z (%) = 192 (14) [M+H]$^+$, 191 (100) [M]$^+$, 150 (22) [M–C$_3$H$_5$]$^+$, 125 (65) [M–C$_5$H$_7$+H]$^+$, 124 (62) [M–C$_5$H$_7$]$^+$, 106 (49) [M–C$_5$H$_8$–NH$_2$]$^+$; HRMS (EI, 40 °C): calc. for C$_{11}$H$_{13}$N^{32}S [M]$^+$ 191.0763; found 191.0764.

1-(3-Chlorophenylthio)-bicyclo[1.1.1]pentane (**104i**)

According to **GP1** 3-chlorothiophenol (**110i**, 51.0 mg, 353 µmol, 1.00 equiv.) was reacted with **1** (590 µL, 352 µmol, 1.00 equiv.) in Et$_2$O (350 µL) for 15 min. The product was obtained as a yellow oil (53 mg, 252 µmol, 72%).

^{1}H NMR (400 MHz, CDCl$_3$): δ = 7.44–7.43 (m, 1H, Ar-H), 7.32–7.30 (m, 1H, Ar-H), 7.24–7.22 (m, 2H, Ar-H), 2.75 (s, 1H, CH), 1.98 (s, 6H, 3 × CH$_2$) ppm; ^{13}C NMR (100 MHz, CDCl$_3$) δ = 136.4 (C$_q$, C_{Ar}Cl), 134.4 (C$_q$, C_{Ar}S), 132.9 (+, CH$_{Ar}$), 131.3 (+, CH$_{Ar}$), 129.9 (+, CH$_{Ar}$), 127.6 (+, CH$_{Ar}$), 54.2 (–, 3 × CH$_2$), 45.5 (C$_q$, C_{Ar}SC), 29.0 (+, CH) ppm; IR (ATR): $\tilde{v}$ = 3467, 2924, 2853, 1746, 1575, 1461, 1378, 1358, 1247, 1206, 1130, 1089, 1069, 1006, 939, 891, 871, 853, 833, 814, 772, 681, 667, 545, 524 cm^{-1}; MS (EI, 20 °C): m/z (%) = 212/210 (15/38) [M]$^+$, 169 (23) [M–C$_3$H$_5$]$^+$, 144 (43) [M–C$_5$H$_7$+H]$^+$, 134 (37) [M–C$_3$H$_5$–Cl]$^+$, 108 (36) [M–C$_5$H$_7$–Cl]$^+$, 67 (100) [C$_5$H$_7$]$^+$; HRMS (EI, 20 °C): calc. for C$_{11}$H$_{11}$^{35}Cl^{32}S [M]$^+$ 210.0270; found 210.0271.

1-(3-Aminophenylthio)-bicyclo[1.1.1]pentane (**104j**)

According to **GP1** 3-aminothiophenol (**110j**, 49.0 mg, 391 µmol, 1.01 equiv.) was reacted with **1** (1.00 mL, 386 µmol, 1.00 equiv.) in Et$_2$O (390 µL) for 15 min. The product was obtained as a pale yellow oil (47 mg, 246 µmol, 64%).

^{1}H NMR (400 MHz, CDCl$_3$): δ = 7.10–7.05 (m, 1H, Ar-H), 6.86–6.81 (m, 1H, Ar-H), 6.78–6.76 (m, 1H, Ar-H), 6.61–6.57 (m, 1H, Ar-H), 3.65 (b, 2H, NH$_2$), 2.72 (s, 1H, CH), 1.97 (s, 6H, 3 × CH$_2$) ppm; ^{13}C NMR (100 MHz, CDCl$_3$): δ = 146.7 (C$_q$, C_{Ar}NH$_2$), 135.0 (C$_q$, C_{Ar}S), 129.6 (+, CH$_{Ar}$), 123.6 (+, CH$_{Ar}$), 119.8 (+, CH$_{Ar}$), 114.4 (+, CH$_{Ar}$), 54.2 (–, 3 × CH$_2$), 45.6 (C$_q$, C_{Ar}SC), 28.8 (+, CH) ppm; IR (ATR): $\tilde{v}$ = 3351, 2977, 2907, 2873, 1616, 1588, 1478, 1438, 1297, 1263, 1204, 1163, 1127, 1078, 992, 886, 771, 687, 529, 446 cm^{-1}; MS (EI, 20 °C): m/z (%) = 192 (6)

[M+H]⁺, 191 (49) [M]⁺, 150 (26) [M–C₃H₅]⁺, 125 (100) [M–C₅H₇+H]⁺, 124 (13) [M–C₅H₇]⁺, 106 (34) [M–C₅H₈–NH₂]⁺; HRMS (EI, 20 °C): calc. for $C_{11}H_{13}N^{32}S$ [M]⁺ 191.0763; found 191.0763.

1-(3-Carboxyphenylthio)-bicyclo[1.1.1]pentane (104k)

According to **GP1** 3-mercaptobenzoic acid (**104k**, 84.0 mg, 545 μmol, 1.00 equiv.) was reacted with **1** (2.00 mL, 544 μmol, 1.00 equiv.) without additional solvent for 1 h. The reaction mixture was not washed with NaOH solution, but directly evaporated instead. The crude product was purified *via* column chromatography (*chex*/EtOAc/AcOH 5:1:0.01). The product was obtained as a colorless solid (63 mg, 286 μmol, 53%).

R_f (chex/EtOAc/AcOH, 5:1:0.01): 0.29; m.p. 108–110 °C; ¹H NMR (400 MHz, CDCl₃): δ = 8.12 (dd, ^4J = 1.8 Hz, ^4J = 1.8 Hz, 1H, Ar-H), 7.94 (ddd, ^3J = 7.8 Hz, ^4J = 1.8 Hz, ^4J = 1.3 Hz, 1H, Ar-H), 7.60 (ddd, ^3J = 7.7 Hz, ^4J = 1.8 Hz, ^4J = 1.3 Hz, 1H, Ar-H), 7.35 (dd, ^3J = 7.8 Hz, ^3J = 7.7 Hz, 1H, Ar-H), 2.69 (s, 1H, C*H*), 1.93 (s, 6H, 3 × C*H₂*) ppm; ¹³C NMR (100 MHz, CDCl₃): δ = 170.5 (C_q, *C*O₂H), 138.4 (+, *C*H_Ar), 135.5 (C_q, *C*_ArS), 134.7 (+, *C*H_Ar), 129.9 (C_q, *C*_ArCO₂H), 129.1 (+, *C*H_Ar), 129.0 (+, *C*H_Ar), 54.3 (–, 3 × *C*H₂), 45.5 (C_q, *C*_ArS*C*), 29.1 (+, *C*H) ppm; IR (ATR): $\tilde{v}$ = 2986, 2906, 2873, 2544, 1686, 1591, 1570, 1474, 1420, 1290, 1260, 1204, 1165, 1125, 1071, 928, 905, 886, 848, 813, 746, 720, 677, 656, 548, 515, 414 cm⁻¹; MS (EI, 60 °C): *m/z* (%) = 221 (5) [M+H]⁺, 220 (36) [M]⁺, 179 (15) [M–C₃H₅]⁺, 154 (35) [M–C₅H₇+H]⁺, 135 (28) [M–C₃H₅–CO₂]⁺, 109 (10) [M–C₅H₇–CO₂]⁺, 85 (20) [M–CO₂H–C₆H₅–CH₂]⁺, 67 (100) [C₅H₇]⁺; HRMS (EI, 60 °C): calc. for $C_{12}H_{12}O_2{}^{32}S$ [M]⁺ 220.0558; found 220.0558.

The acidic CO₂H signal was not visible in ¹H NMR due to peak broadening. Crystallographic information of the product can be found in chapter 7.4.

1-(2-Chlorophenylthio)-bicyclo[1.1.1]pentane (104l)

According to **GP1** 2-chlorothiophenol (**110l**, 39.5 mg, 273 μmol, 1.00 equiv.) was reacted with **1** (1.00 mL, 272 μmol, 1.00 equiv.) in Et₂O (270 μL) for 15 min. The product was obtained as a yellow oil (33 mg, 157 μmol, 58%).

¹H NMR (400 MHz, CDCl₃): δ = 7.57–7.55 (m, 1H, Ar-H), 7.43–7.40 (m, 1H, Ar-H), 7.22–7.19 (m, 2H, Ar-H), 2.75 (s, 1H, C*H*), 2.02 (s, 6H, 3 × C*H₂*) ppm; ¹³C NMR (100 MHz, CDCl₃): δ = 137.1 (C_q, *C*_ArCl), 134.8 (+, *C*H_Ar), 133.9 (C_q, *C*_ArS), 130.0 (+, *C*H_Ar), 128.6 (+, *C*H_Ar), 127.0 (+, *C*H_Ar), 54.4 (–, 3 × *C*H₂), 45.4 (C_q, *C*_ArS*C*), 29.3 (+, *C*H) ppm; IR (ATR): $\tilde{v}$ = 2959, 2913, 2874, 1574, 1449, 1427, 1375, 1249, 1205, 1123, 1035, 923, 894, 748, 658, 551, 463, 432 cm⁻¹;

MS (EI, 20 °C): m/z (%) = 212/210 (10/27) [M]$^+$, 169 (26) [M–C$_3$H$_5$]$^+$, 144 (67) [M–C$_5$H$_7$+H]$^+$, 134 (40) [M–C$_3$H$_5$–Cl]$^+$, 108 (40) [M–C$_5$H$_7$–Cl]$^+$, 67 (100) [C$_5$H$_7$]$^+$; HRMS (EI, 20 °C): calc. for C$_{11}$H$_{11}$^{35}Cl^{32}S [M]$^+$ 210.0270; found 210.0270.

1-(2-Methylphenylthio)-bicyclo[1.1.1]pentane (**104m**)

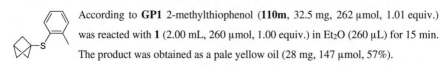

According to **GP1** 2-methylthiophenol (**110m**, 32.5 mg, 262 µmol, 1.01 equiv.) was reacted with **1** (2.00 mL, 260 µmol, 1.00 equiv.) in Et$_2$O (260 µL) for 15 min. The product was obtained as a pale yellow oil (28 mg, 147 µmol, 57%).

^{1}H NMR (400 MHz, CDCl$_3$): δ = 7.48–7.46 (m, 1H, Ar-H), 7.24–7.12 (m, 3H, Ar-H), 2.70 (s, 1H, CH), 2.43 (s, 3H, CH_3), 1.94 (s, 6H, 3 × CH_2) ppm; ^{13}C NMR (100 MHz, CDCl$_3$): δ = 141.0 (C$_q$, C_{Ar}CH$_3$), 135.0 (+, CH$_{Ar}$), 133.5 (C$_q$, C_{Ar}S), 130.4 (+, CH$_{Ar}$), 127.9 (+, CH$_{Ar}$), 126.3 (+, CH$_{Ar}$), 54.2 (–, 3 × CH$_2$), 45.8 (C$_q$, C_{Ar}SCC), 29.0 (+, CH), 21.3 (+, CH$_3$) ppm; IR (ATR): $\tilde{v}$ = 2977, 2909, 2875, 1589, 1469, 1378, 1203, 1127, 1061, 1035, 923, 896, 779, 746, 712, 678, 556, 460, 424 cm^{-1}; MS (EI, 20 °C): m/z (%) = 190 (10) [M]$^+$, 149 (100) [M–C$_3$H$_5$]$^+$, 124 (17) [M–C$_5$H$_7$+H]$^+$, 91 (30) [M–C$_5$H$_7$S]$^+$, 67 (18) [C$_5$H$_7$]$^+$; HRMS (EI, 20 °C): calc. for C$_{12}$H$_{14}$^{32}S [M]$^+$ 190.0811; found 190.0809.

1-(2-Carboxyphenylthio)-bicyclo[1.1.1]pentane (**104o**)

According to **GP1** 2-mercaptobenzoic acid (**110o**, 68.0 mg, 441 µmol, 1.01 equiv.) was reacted with **1** (1.00 mL, 438 µmol, 1.00 equiv.) without additional solvent for 1 h. The reaction mixture was not washed with NaOH solution, but directly evaporated instead. The crude product was purified *via* column chromatography (chex/EtOAc/AcOH 5:1:0.01). The product was obtained as a colorless solid (29 mg, 132 µmol, 30%).

R_f (chex/EtOAc/AcOH, 5:1:0.01): 0.22; m.p. 125–127 °C; ^{1}H NMR (400 MHz, CDCl$_3$): δ = 8.17 (dd, ^3J = 7.8 Hz, ^4J = 1.5 Hz, 1H, Ar-H), 7.64 (dd, ^3J = 7.8 Hz, ^4J = 1.1 Hz, 1H, Ar-H), 7.50 (ddd, ^3J = 7.8 Hz, ^3J = 7.6 Hz, ^4J = 1.5 Hz, 1H, Ar-H), 7.34 (ddd, ^3J = 7.8 Hz, ^3J = 7.6 Hz, ^3J = 1.1 Hz, 1H, Ar-H), 2.81 (s, 1H, CH), 2.08 (s, 6H, 3 × CH_2) ppm; ^{13}C NMR (100 MHz, CDCl$_3$): δ = 169.7 (C$_q$, CO$_2$H), 132.9 (C$_q$, C_{Ar}S), 132.8 (+, 2 × CH$_{Ar}$), 132.7 (+, CH$_{Ar}$), 129.9 (C$_q$, C_{Ar}CO$_2$H), 127.0 (+, CH$_{Ar}$), 54.4 (–, 3 × CH$_2$), 45.3 (C$_q$, C_{Ar}SC), 29.8 (+, CH) ppm; IR (ATR): $\tilde{v}$ = 2971, 2915, 2879, 2643, 1673, 1587, 1560, 1463, 1434, 1406, 1309, 1251, 1206, 1135, 1056, 1043, 917, 893, 806, 737, 692, 650, 557, 491, 464, 416 cm^{-1}; MS (EI, 70 °C): m/z (%) = 221 (1) [M+H]$^+$, 220 (4) [M]$^+$, 179 (10) [M–C$_3$H$_5$]$^+$, 136 (100) [M–C$_3$H$_5$–CO$_2$+H]$^+$, 109 (11) [M–C$_5$H$_7$–CO$_2$]$^+$, 99 (64) [M–

$C_6H_4CO_2H]^+$, 67 (37) $[C_5H_7]^+$; HRMS (EI, 70 °C): calc. for $C_{12}H_{12}O_2{}^{32}S$ [M]$^+$ 220.0558; found 220.0559.

The acidic CO_2H signal was not visible in ^1H NMR due to peak broadening. Crystallographic information of the product can be found in chapter 7.4.

1-(2,6-Dichlorophenylthio)-bicyclo[1.1.1]pentane (**104p**)

According to **GP1** 2,6-dichlorothiophenol (**110p**, 173 mg, 966 µmol, 1.01 equiv.) was reacted with **1** (2.50 mL, 965 µmol, 1.00 equiv.) without additional solvent for 15 min. The product was obtained as a yellow solid (154 mg, 628 µmol, 65%).

m.p. 40–42 °C; ^1H NMR (400 MHz, CDCl$_3$) δ = 7.41 (d, ^3J = 7.7 Hz, 2H, Ar-H), 7.18 (dd, ^3J = 7.7 Hz, ^3J = 7.7 Hz, 1H, Ar-H), 2.67 (s, 1H, CH), 1.96 (s, 6H, 3 × CH_2) ppm; ^{13}C NMR (100 MHz, CDCl$_3$) δ = 142.5 (C$_q$, 2 × C$_{Ar}$Cl), 131.9 (C$_q$, C$_{Ar}$S), 130.2 (+, CH_{Ar}), 128.6 (+, 2 × CH_{Ar}), 54.7 (–, 3 × CH_2), 46.0 (C$_q$, C$_{Ar}$SC), 28.5 (+, CH) ppm; IR (ATR): $\tilde{v}$ = 2980, 2909, 2875, 1553, 1422, 1400, 1202, 1185, 1123, 1086, 926, 892, 773, 708, 557, 557, 517, 476, 436, 398 cm^{-1}; MS (EI, 20 °C): m/z (%) = 248/246/244 (1/5/7) [M]$^+$, 209 (14) [M–Cl]$^+$, 178 (29) [M–C$_5$H$_7$+H]$^+$, 142 (28) [M–C$_5$H$_7$–Cl]$^+$, 107 (10) [M–C$_5$H$_7$–Cl$_2$]$^+$, 67 (100) [M–C$_6$H$_3$Cl$_2$S]$^+$; HRMS (EI, 20 °C): calc. for $C_{11}H_{10}{}^{35}Cl_2{}^{32}S$ [M]$^+$ 243.9875; found 243.9876.

1-(3,5-Dichlorophenylthio)-bicyclo[1.1.1]pentane (**104q**)

According to **GP1** 3,5-dichlorothiophenol (**110q**, 48.5 mg, 271 µmol, 1.01 equiv.) was reacted with **1** (1.00 mL, 267 µmol, 1.00 equiv.) in Et$_2$O (270 µL) for 15 min. The product was obtained as a yellow oil (39 mg, 159 µmol, 60%).

^1H NMR (400 MHz, CDCl$_3$) δ = 7.30 (d, ^3J = 1.8 Hz, 2H, Ar-H), 7.25 (t, ^3J = 1.8 Hz, 1H, Ar-H), 2.78 (s, 1H, CH), 2.02 (s, 6H, 3 × CH_2) ppm; ^{13}C NMR (100 MHz, CDCl$_3$) δ = 138.1 (C$_q$, C$_{Ar}$S), 134.9 (C$_q$, 2 × C$_{Ar}$Cl), 130.7 (+, 2 × CH_{Ar}), 127.5 (+, CH_{Ar}), 54.4 (–, 3 × CH_2), 45.2 (C$_q$, C$_{Ar}$SC), 29.3 (+, CH) ppm; IR (ATR): $\tilde{v}$ = 2981, 2908, 2876, 1554, 1401, 1379, 1206, 1141, 1125, 1100, 894, 854, 795, 670, 428 cm^{-1}; MS (EI, 20 °C): m/z (%) = 248/246/244 (2/10/15) [M]$^+$, 178 (20) [M–C$_5$H$_7$+H]$^+$, 142 (47) [M–C$_5$H$_7$–Cl]$^+$, 107 (12) [M–C$_5$H$_7$–Cl$_2$]$^+$, 99 (36) [C$_5$H$_7$S]$^+$, 67 (100) [M–C$_6$H$_3$Cl$_2$S]$^+$; HRMS (EI, 20 °C): calc. for $C_{11}H_{10}{}^{35}Cl_2{}^{32}S$ [M]$^+$ 243.9875; found 243.9874.

1-(2,4,6-Trimethylphenylthio)-bicyclo[1.1.1]pentane (**104r**)

According to **GP1** 2,4,6-trimethylthiophenol (**110r**, 41.5 mg, 273 μmol, 1.00 equiv.) was reacted with **1** (1.00 mL, 272 μmol, 1.00 equiv.) in Et$_2$O (270 μL) for 15 min. The product was obtained as a pale yellow oil (50 mg, 229 μmol, 84%).

^{1}H NMR (400 MHz, CDCl$_3$): δ = 6.95–6.92 (m, 2H, Ar-H), 2.62 (s, 1H, C*H*), 2.46 (s, 6H, C^{2}C*H$_3$* + C^{6}C*H$_3$*), 2.27 (s, 3H, C^{4}C*H$_3$*), 1.86 (s, 6H, 3 × C*H$_2$*) ppm; ^{13}C NMR (100 MHz, CDCl$_3$): δ = 143.6 (C$_q$, 2 × *C*$_{Ar}$CH$_3$), 138.1 (C$_q$, *C*$_{Ar}$CH$_3$), 128.9 (+, 2 × *C*H$_{Ar}$), 128.6 (C$_q$, C^1), 54.2 (–, 3 × *C*H$_2$), 46.4 (C$_q$, C^{1}S*C*), 28.3 (+, *C*H), 22.4 (+, C^2*C*H$_3$ + C^6*C*H$_3$), 21.2 (+, C^4*C*H$_3$) ppm; IR (ATR): $\tilde{v}$ = 2974, 2908, 2872, 1601, 1448, 1373, 1203, 1128, 1060, 1030, 894, 848, 718, 562, 412 cm^{-1}; MS (EI, 30 °C): *m/z* (%) = 218 (5) [M]$^+$, 203 (11) [M–CH$_3$]$^+$, 177 (100) [M–C$_3$H$_5$]$^+$, 162 (35) [M–C$_3$H$_5$–CH$_3$]$^+$, 119 (21) [M–C$_5$H$_7$S]$^+$; HRMS (EI, 30 °C): calc. for C$_{14}$H$_{18}$^{32}S [M]$^+$ 218.1124; found 218.1123.

5-((Bicyclo[1.1.0]butan-1-ylmethyl)thio)-1-phenyl-1*H*-tetrazole (**117**)

According to **GP1** 5-mercapto-1-phenyltetrazole (**116**, 214 mg, 1.20 mmol, 1.28 equiv.) was reacted with **1** (4.00 mL, 939 μmol, 1.00 equiv.) in THF (1.20 mL) for 15 min. The crude product was purified *via* column chromatography (*n*-pentane/Et$_2$O 25:1). The product was obtained as a yellow oil (14 mg, 57.3 μmol, 6%).

R_f (*n*-pentane/Et$_2$O, 25:1): 0.42; ^{1}H NMR (500 MHz, CDCl$_3$): δ = 8.00–7.96 (m, 2H, Ar-H), 7.59–7.53 (m, 2H, Ar-H), 7.53–7.48 (m, 1H, Ar-H), 4.90 (s, 2H, SC*H$_2$*), 1.79 (d, ^3J = 3.0 Hz, 2H, 2 × CC*H$_a$*H$_b$CH), 1.64 – 1.61 (m, 1H, C*H*), 0.88 (s, 2H, 2 × CCH$_a$*H$_b$*CH) ppm; ^{13}C NMR (125 MHz, CDCl$_3$): δ = 163.5 (C$_q$, S*C*), 135.0 (C$_q$, *C*$_{Ar}$), 129.7 (+, *C*H$_{Ar}$), 129.4 (+, 2 × *C*H$_{Ar}$), 123.9 (+, 2 × *C*H$_{Ar}$), 49.9 (–, S*C*H$_2$), 34.5 (–, 2 × C*C*H$_2$CH), 8.0 (C$_q$, *C*CH$_2$CH), 2.2 (+, *C*H) ppm; MS (GC-MS, BR80): *m/z* = 244, 175, 135, 99, 77 @ 12.15 min. Due to decomposition of the product no HRMS and IR spectra were recorded.

1-(Phenylseleno)-bicyclo[1.1.1]pentane (**130**)

According to **GP1** benzeneselenol (**131**, 64 μL, 94.7 mg, 603 μmol, 1.01 equiv.) was reacted with **1** (1.00 mL, 596 μmol, 1.00 equiv.) in Et$_2$O (600 μL) for 15 min. The product was obtained as a yellow oil (133 mg, 596 μmol, quant.).

^{1}H NMR (400 MHz, CDCl$_3$): δ = 7.57–7.54 (m, 2H, Ar-H), 7.30–7.25 (m, 3H, Ar-H), 2.96 (s, 1H, C*H*), 2.00 (s, 6H, 3 × C*H*$_2$) ppm; ^{13}C NMR (100 MHz, CDCl$_3$): δ = 135.4 (+, 2 × *C*H$_{Ar}$), 131.7 (C$_q$, *C*$_{Ar}$), 128.9 (+, 2 × *C*H$_{Ar}$), 127.6 (+, *C*H$_{Ar}$), 55.4 (–, 3 × *C*H$_2$), 38.9 (C$_q$, *C*$_{Ar}$Se*C*), 31.0 (+, *C*H) ppm; IR (ATR): $\tilde{v}$ = 3056, 2962, 2909, 2874, 1577, 1475, 1436, 1299, 1204, 1117, 1072, 1021, 999, 883, 735, 690, 671, 471 cm^{-1}; MS (EI, 20 °C): *m/z* (%) = 224/222/220 (18/19/5) [M]$^+$, 158 (21) [M–C$_5$H$_7$+H]$^+$, 157 (20) [M–C$_5$H$_7$]$^+$, 78 (33) [C$_6$H$_5$+H]$^+$, 77 (41) [C$_6$H$_5$]$^+$, 67 (100) [C$_5$H$_7$]$^+$; HRMS (EI, 20 °C): calc. for C$_{11}$H$_{12}$^{80}Se [M]$^+$ 224.0104; found 224.0104.

1-(*n*-Propylthio)-bicyclo[1.1.1]pentane (**105a**)

 According to **GP1** 1-propanethiol (**132a**, 34 µL, 27.9 mg, 366 µmol, 1.01 equiv.) was reacted with **1** (1.00 mL, 363 µmol, 1.00 equiv.) in Et$_2$O (360 µL) for 15 min. The product was obtained as a volatile, pale yellow liquid (42 mg, 295 µmol, 81%).

^{1}H NMR (400 MHz, CDCl$_3$): δ = 2.72 (s, 1H, C*H*), 2.51 (t, ^3J = 7.3 Hz, 2H, C*H*$_2$CH$_2$CH$_3$), 1.96 (s, 6H, 3 × CC*H*$_2$CH), 1.60 (td, ^3J = 7.4 Hz, ^3J = 7.3 Hz, 2H, CH$_2$C*H*$_2$CH$_3$), 0.98 (t, ^3J = 7.4 Hz, 3H, C*H*$_3$) ppm; ^{13}C NMR (100 MHz, CDCl$_3$): δ = 53.9 (–, 3 × C*C*H$_2$CH), 44.7 (C$_q$, *C*CH$_2$CH), 33.3 (–, *C*H$_2$CH$_2$CH$_3$), 28.8 (+, *C*H), 23.9 (–, CH$_2$*C*H$_2$CH$_3$), 13.7 (+, *C*H$_3$) ppm; IR (ATR): $\tilde{v}$ = 2961, 2907, 2872, 1450, 1376, 1291, 1205, 1140, 902, 797 cm^{-1}; MS (GC-MS, BR80): *m/z* = 141, 100, 99, 85, 67 @ 5.14 min. Due to the nonpolar and volatile nature of this compound no HRMS measurement was possible with EI, FAB or ESI.

1-(*iso*-Propylthio)-bicyclo[1.1.1]pentane (**105b**)

According to **GP1** 2-propanethiol (**132b**, 34 µL, 27.9 mg, 366 µmol, 1.03 equiv.) was reacted with **1** (1.00 mL, 357 µmol, 1.00 equiv.) in Et$_2$O (360 µL) for 15 min. The product was obtained as a volatile, colorless liquid (18 mg, 127 µmol, 35%).

^{1}H NMR (400 MHz, CDCl$_3$): δ = 2.96 (sept, ^3J = 6.8 Hz, 1H, C*H*(CH$_3$)$_2$), 2.71 (s, 1H, C*H*(CH$_2$)$_3$), 2.00 (s, 6H, 3 × C*H*$_2$), 1.28 (d, ^3J = 6.8 Hz, 6H, 2 × C*H*$_3$) ppm; ^{13}C NMR (100 MHz, CDCl$_3$): δ = 54.6 (–, 3 × *C*H$_2$), 44.5 (C$_q$, CHS*C*), 35.8 (+, *C*H(CH$_3$)$_2$), 29.2 (+, *C*H(CH$_2$)$_3$), 24.8 (+, 2 × *C*H$_3$) ppm; IR (ATR): $\tilde{v}$ = 2974, 2908, 2873, 1447, 1381, 1364, 1243, 1206, 1139, 1050, 903, 646 cm^{-1}; MS (GC-MS, BR80): *m/z* = 141, 100, 99, 85, 67 @ 4.55 min. Due to the nonpolar and volatile nature of this compound no HRMS measurement was possible with EI, FAB or ESI.

1-(*n*-Butylthio)-bicyclo[1.1.1]pentane (**105c**)

According to **GP1** 1-butanethiol (**132c**, 39 µL, 32.8 mg, 364 µmol, 1.00 equiv.) was reacted with **1** (1.00 mL, 363 µmol, 1.00 equiv.) in Et$_2$O (360 µL) for 15 min. The product was obtained as a volatile, pale yellow liquid (38 mg, 243 µmol, 67%).

^{1}H NMR (400 MHz, CDCl$_3$): δ = 2.72 (s, 1H, C*H*), 2.53 (t, ^3J = 7.4 Hz, 2H, C*H$_2$*CH$_2$CH$_2$CH$_3$), 1.96 (s, 6H, 3 × CC*H$_2$*CH), 1.61–1.52 (m, 2H, CH$_2$CH$_2$C*H$_2$*CH$_3$), 1.44–1.35 (m, 2H, CH$_2$C*H$_2$*CH$_2$CH$_3$), 0.91 (t, ^3J = 7.3 Hz, 3H, C*H$_3$*) ppm; ^{13}C NMR (100 MHz, CDCl$_3$): δ = 53.9 (–, 3 × C*C*H$_2$CH), 44.7 (C$_q$, *C*CH$_2$CH), 33.3 (–, CH$_2$CH$_2$*C*H$_2$CH$_3$), 31.0 (–, *C*H$_2$CH$_2$CH$_2$CH$_3$), 28.8 (+, *C*H), 22.2 (–, CH$_2$*C*H$_2$CH$_2$CH$_3$), 13.9 (+, *C*H$_3$) ppm; IR (ATR): $\tilde{v}$ = 2959, 2907, 2872, 1458, 1377, 1274, 1205, 1141, 903, 745, 395 cm^{-1}; MS (GC-MS, BR80): *m/z* = 155, 115, 99, 85, 67 @ 6.20 min. Due to the nonpolar and volatile nature of this compound no HRMS measurement was possible with EI, FAB or ESI.

1-(*tert*-Butylthio)-bicyclo[1.1.1]pentane (**105d**)

According to **GP1** 2-methyl-2-propanethiol (**132d**, 36 µL, 28.8 mg, 319 µmol, 1.02 equiv.) was reacted with **1** (1.00 mL, 312 µmol, 1.00 equiv.) in Et$_2$O (320 µL) for 15 min. The product was obtained as a volatile, pale yellow liquid (33 mg, 211 µmol, 68%).

^{1}H NMR (400 MHz, CDCl$_3$): δ = 2.69 (s, 1H, C*H*), 2.08 (s, 6H, 3 × C*H$_2$*), 1.36 (s, 9H, 3 × C*H$_3$*) ppm; ^{13}C NMR (100 MHz, CDCl$_3$): δ = 56.1 (–, 3 × C*H$_2$*), 44.8 (C$_q$, *C*SC(CH$_3$)$_3$), 44.1 (C$_q$, *C*(CH$_3$)$_3$), 32.0 (+, 3 × C*H$_3$*), 30.7 (+, C*H*) ppm; IR (ATR): $\tilde{v}$ = 2962, 2910, 2874, 1456, 1362, 1260, 1208, 1164, 1135, 903, 803, 597 cm^{-1}; MS (GC-MS, BR80): *m/z* = 156, 100, 85, 67 @ 5.18 min. Due to the nonpolar and volatile nature of this compound no HRMS measurement was possible with EI, FAB or ESI.

1-(2-Methyl-2-undecanethio)-bicyclo[1.1.1]pentane (**105e**)

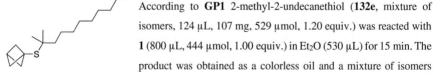

According to **GP1** 2-methyl-2-undecanethiol (**132e**, mixture of isomers, 124 µL, 107 mg, 529 µmol, 1.20 equiv.) was reacted with **1** (800 µL, 444 µmol, 1.00 equiv.) in Et$_2$O (530 µL) for 15 min. The product was obtained as a colorless oil and a mixture of isomers (118 mg, 439 µmol, 99%).

MS (EI, 20 °C): m/z (%) = 268 (1) [M]$^+$, 227 (1) [M-C$_3$H$_5$]$^+$, 169 (5) [M-C$_5$H$_7$S]$^+$, 113 (11) [M-C$_9$H$_{15}$S]$^+$, 99 (23) [M-C$_{12}$H$_{25}$]$^+$, 85 (55) [M-C$_{11}$H$_{19}$S]$^+$, 71 (68) [M-C$_{12}$H$_{21}$S]$^+$, 67 (11) [C$_5$H$_7$]$^+$, 57 (100) [M-C$_{13}$H$_{23}$S]$^+$; HRMS (EI, 20 °C): calc. for C$_{17}$H$_{32}$^{32}S [M]$^+$ 268.2225; found 268.2224.

The NMR signals could not be assigned to the mixture of isomers.

1-(Benzylthio)-bicyclo[1.1.1]pentane (**105f**)

According to **GP1** benzyl mercaptane (**132f**, 110 µL, 116 mg, 937 µmol, 1.00 equiv.) was reacted with **1** (3.00 mL, 936 µmol, 1.00 equiv.) in Et$_2$O (940 µL) for 15 min. The product was obtained as a pale yellow oil (94 mg, 494 µmol, 53%).

^{1}H NMR (400 MHz, CDCl$_3$): δ = 7.32–7.17 (m, 5H, Ar-H), 3.73 (s, 2H, C$_6$H$_5$CH_2S), 2.64 (s, 1H, CH), 1.68 (s, 6H, 3 × CCH_2CH) ppm; ^{13}C NMR (100 MHz, CDCl$_3$): δ = 138.9 (C$_q$, C_{Ar}), 128.9 (+, 2 × CH$_{Ar}$), 128.5 (+, 2 × CH$_{Ar}$), 127.0 (+, CH$_{Ar}$), 53.8 (−, 3 × CCH_2CH), 44.8 (C$_q$, SC), 35.9 (−, C$_{Ar}$$CH_2$S), 29.0 (+, CH) ppm; IR (ATR): $\tilde{v}$ = 3060, 3026, 2974, 2906, 2872, 1600, 1493, 1451, 1205, 1137, 1068, 1028, 903, 862, 760, 696, 563, 472, 440 cm^{-1}; MS (EI, 20 °C): m/z (%) = 190 (1) [M]$^+$, 99 (4) [M–C$_6$H$_5$CH$_2$]$^+$, 91 (100) [M–C$_5$H$_7$S]$^+$, 67 (3) [C$_5$H$_7$]$^+$; HRMS (EI, 20 °C): calc. for C$_{12}$H$_{14}$^{32}S [M]$^+$ 190.0811; found 190.0811.

1-(2-(9-Fluorenylmethoxycarbonylamino)ethylthio)-bicyclo[1.1.1]pentane (**105h**)

According to **GP1** Fmoc-cysteamine (**132h**, 300 mg, 1.00 mmol, 1.00 equiv.) was reacted with **1** (3.72 mL, 1.20 mmol, 1.20 equiv.) in THF (3.00 mL) for 3 h. The reaction mixture was not washed with NaOH solution, but directly evaporated instead. The crude product was purified *via* column chromatography (chex/EtOAc 10:1). The product was obtained as a colorless solid (180 mg, 493 µmol, 49%).

R_f (chex/EtOAc, 10:1): 0.15; m.p. 88–90 °C; ^{1}H NMR (400 MHz, CDCl$_3$): δ = 7.77 (dt, ^3J = 7.6 Hz, ^4J = 0.9 Hz, 2H, Ar-H), 7.60 (dd, ^3J = 7.5 Hz, ^4J = 1.1 Hz, 2H, Ar-H), 7.43 – 7.37 (m, 2H, Ar-H), 7.32 (td, ^3J = 7.4 Hz, ^4J = 1.2 Hz, 2H, Ar-H), 5.14 (bt, ^3J = 6.5 Hz, 1H, NH), 4.42 (d, ^3J = 6.9 Hz, 2H, CO$_2$CH_2), 4.22 (t, ^3J = 6.9 Hz, 1H, CO$_2$CH$_2$CH), 3.38 (td, ^3J = 6.6 Hz, ^3J = 6.5 Hz, 2H, SCH$_2$CH_2), 2.73 (s, 1H, CCH$_2$CH), 2.68 (t, ^3J = 6.6 Hz, 2H, SCH_2CH$_2$), 1.98 (s, 6H, 3 × CCH_2CH) ppm; ^{13}C NMR (100 MHz, CDCl$_3$): δ = 156.4 (C$_q$, CO_2CH$_2$), 144.1 (C$_q$, 2 × C_{Ar}), 141.5 (C$_q$, 2 × C_{Ar}), 127.8 (+, 2 × CH$_{Ar}$), 127.2 (+, 2 × CH$_{Ar}$), 125.2 (+, 2 × CH$_{Ar}$), 120.1 (+,

$2 \times CH_{Ar}$), 66.8 (–, CO_2CH_2), 54.0 (–, $3 \times CCH_2CH$), 47.4 (+, CO_2CH_2CH), 44.3 (C_q, $CSCH_2$), 41.3 (–, SCH_2CH_2), 31.5 (–, SCH_2CH_2), 28.8 (+, CH) ppm; IR (ATR): $\tilde{v}$ = 3364, 2979, 2962, 2924, 2910, 2876, 1691, 1524, 1463, 1445, 1432, 1309, 1258, 1242, 1203, 1137, 1102, 1082, 1048, 1030, 1004, 989, 970, 939, 898, 778, 758, 734, 645, 620, 606, 585, 571, 526, 480, 424, 385 cm^{-1}; MS (EI, 60 °C): m/z (%) = 365 (0.1) [M]$^+$, 178 (100) [$C_{14}H_{10}$]$^+$, 67 (3) [C_5H_7]$^+$; HRMS (EI, 60 °C): calc. for $C_{22}H_{23}NO_2{}^{32}S$ [M]$^+$ 365.1450; found 365.1452.

2-(1-Bicyclo[1.1.1]pentylthio)acetic acid (**105i**)

According to **GP1** mercaptoacetic acid (**132i**, 40 µL, 53.0 mg, 576 µmol, 1.01 equiv.) was reacted with **1** (2.00 mL, 568 µmol, 1.00 equiv.) in Et$_2$O (570 µL) for 15 min. The reaction mixture was not washed with NaOH solution, but directly evaporated instead. The crude product was purified *via* column chromatography (*chex*/EtOAc/TFA 10:1:0.01). The product was obtained as a colorless oil (59 mg, 373 µmol, 66%).

R_f (*chex*/EtOAc/TFA, 10:1:0.01): 0.20; ^{1}H NMR (400 MHz, CDCl$_3$): δ = 3.31 (s, 2H, CSCH_2), 2.75 (s, 1H, CH), 2.01 (s, 6H, $3 \times$ CCH_2CH) ppm; ^{13}C NMR (100 MHz, CDCl$_3$): δ = 175.3 (C_q, CO_2H), 53.6 (–, $3 \times CCH_2CH$), 44.3 (C_q, $CSCH_2$), 33.2 (–, $CSCH_2$), 28.7 (+, CH) ppm; IR (ATR): $\tilde{v}$ = 2979, 2910, 2876, 2670, 1705, 1419, 1291, 1206, 1134, 900, 785, 668, 581, 464 cm^{-1}; MS (EI, 30 °C): m/z (%) = 158 (1) [M]$^+$, 157 (2) [M–H]$^+$, 99 (100) [M–CH$_2$CO$_2$H]$^+$, 67 (79) [C_5H_7]$^+$; HRMS (EI, 30 °C): calc. for $C_7H_{10}O_2{}^{32}S$ [M]$^+$ 158.0402; found 158.0403.

The acidic CO$_2$H signal was not visible in ^{1}H NMR due to peak broadening.

Methyl 3-(1-bicyclo[1.1.1]pentylthio) propionate (**105k**)

According to **GP1** methyl 3-mercaptopropionate (**132k**, 60 µL, 65.1 mg, 542 µmol, 1.22 equiv.) was reacted with **1** (800 µL, 444 µmol, 1.00 equiv.) in Et$_2$O (540 µL) for 15 min. The product was obtained as a colorless oil (82 mg, 440 µmol, 99%).

^{1}H NMR (400 MHz, CDCl$_3$): δ = 3.69 (s, 3H, OCH_3), 2.79 (t, ^3J = 7.6 Hz, 2H, SCH_2CH$_2$), 2.73 (s, 1H, CH), 2.60 (t, ^3J = 7.6 Hz, 2H, SCH$_2$CH_2), 1.97 (s, 6H, $3 \times$ CCH_2CH) ppm; ^{13}C NMR (100 MHz, CDCl$_3$): δ = 172.5 (C_q, CO_2CH_3), 53.9 (–, $3 \times CCH_2CH$), 51.9 (+, OCH_3), 44.4 (C_q, $CSCH_2$), 35.5 (–, SCH_2CH_2), 28.9 (+, CH), 26.2 (–, SCH_2CH_2) ppm; IR (ATR): $\tilde{v}$ = 2976, 2908, 2874, 1737, 1435, 1354, 1299, 1285, 1245, 1207, 1171, 1154, 1139, 1018, 980, 931, 901, 823, 778, 741, 674, 445 cm^{-1}; MS (EI, 20 °C): m/z (%) = 186 (0.5) [M]$^+$, 185 (1) [M–H]$^+$, 155 (5) [M–

CH$_3$O]$^+$, 99 (100) [C$_5$H$_7$S]$^+$, 67 (48) [C$_5$H$_7$]$^+$; HRMS (EI, 20 °C): calc. for C$_9$H$_{14}$O$_2$^{32}S [M–H]$^+$ 185.0636; found 185.0637.

1-(Tri-*iso*-propylsilanethio)-bicyclo[1.1.1]pentane (**105l**)

According to **GP1** tri-*iso*-propylsilanethiol (**132l**, 299 µL, 265 mg, 1.39 mmol, 1.25 equiv.) was reacted with **1** (2.00 mL, 1.11 mmol, 1.00 equiv.) in Et$_2$O (1.39 mL) for 15 min. The product was obtained as a colorless oil (283 mg, 1.10 mmol, 99%).

^{1}H NMR (400 MHz, CDCl$_3$): δ = 2.61 (s, 1H, CCH$_2$C*H*), 2.04 (s, 6H, 3 × CC*H*$_2$CH), 1.29–1.19 (m, 3H, 3 × C*H*(CH$_3$)$_2$), 1.11 (d, ^3J = 7.3 Hz, 18H, 3 × CH(C*H*$_3$)$_2$) ppm; ^{13}C NMR (100 MHz, CDCl$_3$): δ = 57.5 (–, 3 × C*C*H$_2$CH), 41.1 (C$_q$, *C*SSi), 28.9 (+, CC*H*$_2$CH), 18.9 (+, 3 × CH(*C*H$_3$)$_2$), 13.2 (+, 3 × *C*H(CH$_3$)$_2$) ppm; IR (ATR): $\tilde{v}$ = 2963, 2944, 2865, 1462, 1382, 1366, 1208, 1127, 1069, 1017, 993, 919, 880, 671, 645, 591, 568, 521, 493, 452 cm^{-1}; MS (EI, 20 °C): *m/z* (%) = 256 (0.5) [M]$^+$, 213 (12) [M–CH(CH$_3$)$_2$]$^+$, 171 (61) [M–(CH(CH$_3$)$_2$)$_2$+H]$^+$, 143 (89) [M–C$_5$H$_7$S– CH$_3$+H]$^+$, 67 (100) [C$_5$H$_7$]$^+$; HRMS (EI, 20 °C): calc. for C$_{11}$H$_{21}$^{32}S^{28}Si [M–CH(CH$_3$)$_2$]$^+$ 213.1128; found 213.1127.

1-(Bicyclo[1.1.1]pentylthio)-bicyclo[1.1.1]pentane (**134**)

In a quartz flask a 0.9 M hydrogen sulfide solution in THF (**133**, 340 µL, 306 µmol, 1.00 equiv.) was added to **1** (1.10 mL, 611 µmol, 2.00 equiv.) under argon atmosphere. The reaction mixture was irradiated with UV light (254 nm, 500 W) for 15 min at room temperature. Then the reaction mixture was diluted with 10 mL *n*-pentane and washed with 5 mL of a 1 M NaOH solution. The organic phase was dried over Na$_2$SO$_4$ and the solvent was removed under reduced pressure at room temperature to obtain the product as a volatile, pale yellow liquid (51 mg, 307 µmol, quant.).

^{1}H NMR (400 MHz, CDCl$_3$): δ = 2.71 (s, 2H, 2 × C*H*), 2.04 (s, 12H, 6 × C*H*$_2$) ppm; ^{13}C NMR (100 MHz, CDCl$_3$): δ = 55.2 (–, 6 × *C*H$_2$), 44.2 (C$_q$, 2 × S*C*), 30.1 (+, 2 × *C*H) ppm; IR (ATR): $\tilde{v}$ = 2975, 2909, 2874, 1447, 1201, 1131, 894, 471 cm^{-1}; MS (EI, 70 eV): *m/z* (%) = 166 (9) [M]$^+$, 125 (16) [M–C$_3$H$_5$]$^+$, 99 (34) [M–C$_5$H$_7$]$^+$, 85 (54) [M–(C$_3$H$_5$)$_2$+H]$^+$, 67 (100) [C$_5$H$_7$]$^+$; HRMS (EI, 70 eV): calc. for C$_{10}$H$_{14}$^{32}S [M]$^+$ 166.0811; found 166.0812.

Thiophenol-d₁ (**110a**_d_)

Thiophenol (**110a**) (200 µL, 216 mg, 1.96 mmol, 1.00 equiv.) and D_2O (400 µL, 440 mg, 22.0 mmol, 11.2 equiv.) were mixed in a vial and treated with ultrasonification for 2 h. The layers were separated by centrifugation and the organic layer was dried under reduced pressure. The ratio of deuterated thiophenol (**110a**_d_) to thiophenol (**110a**) was determined with ^1H NMR and a deuteration degree of 85% was found. The yield was not determined.

^1H NMR (300 MHz, $CDCl_3$): δ = 7.43–7.21 (m, 5H, Ar-H) ppm.

1-(Phenylthio)-3-d-bicyclo[1.1.1]pentane (**104a**_d_)

According to **GP1** 85% deuterated thiophenol (**110a**_d_, 34 µL, 36.5 mg, 328 µmol, 1.00 equiv.) was reacted with **1** (1.00 mL, 357 µmol, 1.09 equiv.) in Et_2O (330 µL) for 15 min. A mixture of **104a** and **104a**_d_ (15:85, determined by ^1H NMR) was obtained as a pale yellow oil (58 mg, 327 µmol, quant.).

^1H NMR (400 MHz, $CDCl_3$): δ = 7.46–7.43 (m, 2H, Ar-H), 7.33–7.26 (m, 3H, Ar-H), 1.95 (s, 6H, 3 × CH_2) ppm; ^{13}C NMR (100 MHz, $CDCl_3$) δ = 134.2 (C_q, C_{Ar}), 133.6 (+, 2 × CH_{Ar}), 128.9 (+, 2 × CH_{Ar}), 127.6 (+, CH_{Ar}), 54.0 (−, 3 × CH_2), 45.8 (C_q, $C_{Ar}SC$), 28.6 (+, t, ^1J = 26 Hz, CD) ppm; IR (ATR): $\tilde{v}$ = 2980, 2910, 2875, 2227, 1583, 1473, 1438, 1201, 1120, 1088, 1066, 1024, 895, 743, 691, 548, 502, 422 cm^{-1}; MS (EI, 20 °C): m/z (%) = 177 (45) [M]$^+$, 135 (74) [M–$C_3H_4{}^2H$]$^+$, 110 (100) [M–$C_5H_6{}^2H$+H]$^+$, 109 (27) [M–$C_5H_6{}^2H$]$^+$, 100 (11) [M–C_6H_5]$^+$, 77 (14) [C_6H_5]$^+$, 68 (61) [$C_5H_6{}^2H$]$^+$; HRMS (EI, 20 °C): calc. for $C_{11}H_{11}{}^2H^{32}S$ [M]$^+$ 177.0717; found 177.0715.

7.2.3 Insertion of [1.1.1]propellane into disulfide bonds

Bis(2,6-dichlorophenyl) disulfide (**138d**)

2,6-Dichlorothiophenol (2.00 g, 11.2 mmol, 1.00 equiv.) was mixed with dimethyl sulfoxide (794 µL, 872 mg, 11.2 mmol, 1.00 equiv.) and stirred at 80 °C for 16 h. The mixture was cooled to room temperature, CH_2Cl_2 (400 mL) was added and the organic phase was washed with water (3 × 40 mL), dried by the addition of Na_2SO_4, filtered and the solvent was removed under reduced pressure. The product was obtained as a pale yellow solid (1.84 g, 5.18 mmol, 93%).

^1H NMR (300 MHz, $CDCl_3$): δ = 7.34 (dd, ^3J = 8.0 Hz, ^4J = 0.9 Hz, 4H, Ar-H), 7.20 (dd, ^3J = 8.8 Hz, ^3J = 7.1 Hz, 2H, Ar-H) ppm.

The analytical data is in accordance with the literature.[161]

1-Iodo-2-phenylbenzene (**186**)

Tert-butyl nitrite (2.11 mL, 1.83 g, 18.0 mmol, 3.00 equiv.) was added to a solution of *para*-toluenesulfonic acid monohydrate (3.37 g, 18.0 mmol, 3.00 equiv.) in CH$_3$CN (25.0 mL), followed by 2-phenylaniline (1.00 g, 5.90 mmol, 1.00 equiv.). After consumption of the starting material, potassium iodide (4.90 g, 30.0 mmol, 5.00 equiv.), dissolved in 5.0 mL water, was added slowly and the mixture was stirred for 2 h at room temperature. The product was extracted with EtOAc (2 × 50 mL) and washed with 50 mL of sat. Na$_2$S$_2$O$_3$ and 50 mL of water. The aqueous phase was back-extracted with EtOAc (50 mL). The combined organic layers were dried by the addition of Na$_2$SO$_4$. The mixture was filtered through a glass funnel and the solvent was evaporated under reduced pressure. The crude product was purified *via* flash-chromatography (*c*hex) and the product was obtained as a yellow oil (1.38 g, 4.93 mmol, 83%).

R_f (*c*hex): 0.51; ^{1}H NMR (400 MHz, CDCl$_3$): δ = 7.97 (dd, ^3J = 7.9 Hz, ^4J = 1.2 Hz, 1H, Ar-H), 7.48–7.37 (m, 4H, Ar-H), 7.36–7.30 (m, 3H, Ar-H), 7.04 (td, ^3J = 7.6 Hz, ^4J = 1.9 Hz, 1H, Ar-H) ppm; ^{13}C NMR (100 MHz, CDCl$_3$): δ = 146.8 (C$_q$, C_{Ar}C$_{Ar}$), 144.4 (C$_q$, C_{Ar}C$_{Ar}$), 139.6 (+, CH_{Ar}), 130.2 (+, CH_{Ar}), 129.4 (+, 2 × CH_{Ar}), 128.9 (+, CH_{Ar}), 128.2 (+, CH_{Ar}), 128.1 (+, 2 × CH_{Ar}), 127.8 (+, CH_{Ar}), 98.8 (C$_q$, C_{Ar}I) ppm.

The analytical data is in accordance with the literature.[162]

Di-2-biphenylyl disulfide (**138g**)

1-Iodo-2-phenylbenzene (**186**, 1.27 g, 4.50 mmol, 1.00 equiv.), copper(I) iodide (86.4 mg, 453 µmol, 0.100 equiv.), sulfur (145 mg, 4.50 mmol, 1.00 equiv.) and sodium sulfide nonahydrate (1.09 g, 4.50 mmol, 1.00 equiv.) were mixed in a vial under argon atmosphere. DMF (8.50 mL) was added and the reaction mixture was stirred at 100 °C for 16 h. The reaction was quenched by the addition of 1.0 mL of water. The reaction mixture was extracted with EtOAc (2 × 50 mL). The combined organic phase was dried by the addition of Na$_2$SO$_4$, filtered and the solvent was removed under reduced pressure. The crude product was purified *via* flash-chromatography (*c*hex). After trituration with Et$_2$O, the product was obtained as a white solid (0.658 g, 1.78 mmol, 39%).

R_f (chex): 0.22; ^{1}H NMR (400 MHz, CDCl$_3$): δ = 7.60 (dd, ^3J = 6.7 Hz, ^4J = 2.2 Hz, 2H, Ar-H), 7.47–7.37 (m, 10H, Ar-H), 7.28–7.20 (m, 6H, Ar-H) ppm; ^{13}C NMR (100 MHz, CDCl$_3$): δ = 141.4 (C$_q$, 2 × C_{Ar}), 139.8 (C$_q$, 2 × C_{Ar}), 135.1 (C$_q$, 2 × C_{Ar}), 130.2 (+, 2 × CH_{Ar}), 129.7 (+, 4 × CH_{Ar}), 128.4 (+, 2 × CH_{Ar}), 128.4 (+, 4 × CH_{Ar}), 127.9 (+, 2 × CH_{Ar}), 127.3 (+, 2 × CH_{Ar}), 126.7 (+, 2 × CH_{Ar}) ppm.

The analytical data is in accordance with the literature.[163] Crystallographic information of the product can be found in chapter 7.4.

2,2'-Biphenyl disulfide (**138i**)

Biphenyl (5.00 g, 32.4 mmol, 1.00 equiv.) was added in small portions to a solution of tetramethylethylenediamine (9.79 mL, 7.54 g, 64.9 mmol, 2.00 equiv.) in 2.5 M n-butyllithium solution (25.9 mL, 4.15 g, 64.9 mmol, 2.00 equiv.) in n-hexane at –15 °C. After complete addition, the mixture was warmed to room temperature overnight. Sulfur (2.29 g, 71.3 mmol, 2.20 equiv.) was added and the reaction mixture was stirred for 3 h. The mixture was poured into water (300 mL) and extracted with CH$_2$Cl$_2$ (3 × 20 mL). The combined organic phase was dried by the addition of Na$_2$SO$_4$, filtered and the solvent was removed under reduced pressure. The crude product was purified via flash-chromatography (n-pentane). The product was obtained as a yellow solid (300 mg, 1.39 mmol, 4%).

R_f (n-pentane): 0.40; ^{1}H NMR (300 MHz, CDCl$_3$): δ = 7.70 (dd, ^3J = 7.8 Hz, ^4J = 1.5 Hz, 2H, Ar-H), 7.52 (dd, ^3J = 7.6 Hz, ^4J = 1.5 Hz, 2H, Ar-H), 7.38 (ddd, ^3J = 7.8 Hz, ^3J = 7.6 Hz, ^4J = 1.5 Hz, 2H, Ar-H), 7.32 – 7.25 (ddd, ^3J = 7.8 Hz, ^3J = 7.6 Hz, ^4J = 1.5 Hz, 2H, Ar-H) ppm.

The analytical data is in accordance with the literature.[164]

1,3-Bis(phenylthio)bicyclo[1.1.1]pentane (**106a**)

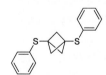

According to **GP2** diphenyl disulfide (**138a**, 342 mg, 1.57 mmol, 3.00 equiv.) was reacted with **1** (1.00 mL, 522 μmol, 1.00 equiv.) in THF (1.00 mL). The crude product was purified via column chromatography (n-pentane to n-pentane/Et$_2$O 100:1). The product was obtained as a colorless solid (145 mg, 510 μmol, 98%).

R_f (n-pentane): 0.19; m.p. 53–55 °C; ^{1}H NMR (400 MHz, CDCl$_3$): δ = 7.42–7.39 (m, 4H, Ar-H), 7.31–7.28 (m, 6H, Ar-H), 2.02 (s, 6H, 3 × CH_2) ppm; ^{13}C NMR (100 MHz, CDCl$_3$): δ = 134.0 (+, 4 × CH_{Ar}), 133.3 (C$_q$, 2 × C_{Ar}S), 129.0 (+, 4 × CH_{Ar}), 128.1 (+, 2 × CH_{Ar}), 57.5 (–, 3 × CH_2), 42.7

(C$_q$, 2 × CCH$_2$) ppm; IR (ATR): $\tilde{v}$ = 3055, 2989, 2961, 2925, 2907, 2867, 1951, 1878, 1715, 1578, 1572, 1472, 1436, 1388, 1327, 1300, 1197, 1159, 1126, 1099, 1088, 1062, 1020, 1000, 987, 928, 907, 890, 849, 759, 739, 703, 687, 554, 507, 476, 422, 408 cm^{-1}; MS (EI, 50 °C): m/z (%) = 284 (3) [M]$^+$, 175 (100) [M–C$_6$H$_5$S]$^+$, 109 (13) [C$_6$H$_5$S]$^+$; HRMS (EI, 50 °C): calc. for C$_{17}$H$_{16}$^{32}S$_2$ [M]$^+$ 284.0693; found 284.0692.

Crystallographic information of the product can be found in chapter 7.4.

1,3-Bis(4-chlorophenylthio)bicyclo[1.1.1]pentane (**106b**)

According to **GP2** bis(4-chlorophenyl) disulfide (**138b**, 411 mg, 1.43 mmol, 3.00 equiv.) was reacted with **1** (1.05 mL, 476 µmol, 1.00 equiv.) in THF (1.00 mL). The crude product was purified *via* column chromatography (*n*-pentane). The product was obtained as a colorless solid (165 mg, 467 µmol, 98%).

R_f (*n*-pentane): 0.38; m.p. 93–95 °C; ^{1}H NMR (400 MHz, CDCl$_3$): δ = 7.35–7.27 (m, 8H, Ar-H), 1.99 (s, 6H, 3 × CH_2) ppm; ^{13}C NMR (100 MHz, CDCl$_3$): δ = 135.4 (+, 4 × CH$_{Ar}$), 134.5 (C$_q$, 2 × C$_{Ar}$Cl), 131.6 (C$_q$, 2 × C$_{Ar}$S), 129.3 (+, 4 × CH$_{Ar}$), 57.4 (+, 3 × CH$_2$), 42.7 (C$_q$, 2 × CCH$_2$) ppm; IR (ATR): $\tilde{v}$ = 2987, 2959, 2924, 2907, 2868, 2854, 1897, 1740, 1639, 1572, 1473, 1443, 1384, 1289, 1262, 1196, 1171, 1133, 1089, 1010, 925, 892, 817, 776, 761, 744, 703, 557, 544, 509, 476, 425, 382 cm^{-1}; MS (EI, 90 °C): m/z (%) = 356/354/352 (1/2/3) [M]$^+$, 211/209 (34/100) [M–C$_6$H$_4$ClS]$^+$, 174 (13) [M–C$_6$H$_4$ClS–Cl]$^+$, 145/143 (5/12) [C$_6$H$_4$ClS]$^+$, 108 (10) [C$_6$H$_4$S]$^+$; HRMS (EI, 90 °C): calc. for C$_{17}$H$_{14}$^{35}Cl$_2$^{32}S$_2$ [M]$^+$ 351.9916; found 351.9914.

1,3-Bis(3,5-dichlorophenylthio)bicyclo[1.1.1]pentane (**106c**)

According to **GP2** bis(3,5-dichlorophenyl) disulfide (**138c**, 513 mg, 1.44 mmol, 3.00 equiv.) was reacted with **1** (1.05 mL, 481 µmol, 1.00 equiv.) in THF (1.00 mL). The crude product was purified *via* column chromatography (*n*-pentane). The product was obtained as a colorless solid (194 mg, 459 µmol, 96%).

R_f (*n*-pentane): 0.72; m.p. 94–96 °C; ^{1}H NMR (400 MHz, CDCl$_3$): δ = 7.30–7.28 (m, 6H, Ar-H), 2.12 (s, 6H, 3 × CH_2) ppm; ^{13}C NMR (100 MHz, CDCl$_3$): δ = 136.6 (C$_q$, 2 × C$_{Ar}$S), 135.2 (C$_q$, 4 × C$_{Ar}$Cl), 131.3 (+, 4 × CH$_{Ar}$), 128.4 (+, 2 × CH$_{Ar}$), 57.8 (+, 3 × CH$_2$), 42.5 (C$_q$, 2 × CCH$_2$) ppm; IR (ATR): $\tilde{v}$ = 3118, 3081, 3054, 3000, 2963, 2915, 2876, 2850, 1742, 1568, 1553, 1503, 1445, 1421, 1401, 1378, 1360, 1288, 1203, 1137, 1125, 1096, 1050, 1018, 990, 926, 894, 874, 837, 790,

660, 561, 530, 521, 507, 463, 428, 409, 390 cm^{-1}; MS (EI, 100 °C): m/z (%) = 424/422/420 (0.6/1/0.5) [M]$^+$, 247/245/243 (13/64/100) [M–C$_6$H$_3$Cl$_2$S]$^+$, 210/208 (13/18) [M–C$_6$H$_3$Cl$_2$S–Cl]$^+$, 179/177 (4/7) [C$_6$H$_3$Cl$_2$S]$^+$, 144/142 (3/8) [C$_6$H$_3$ClS]$^+$; HRMS (EI, 100 °C): calc. for C$_{17}$H$_{12}$^{35}Cl$_4$^{32}S$_2$ [M]$^+$ 419.9134, found 419.9135.

1,3-Bis(2,6-dichlorophenylthio)bicyclo[1.1.1]pentane (106d)

According to **GP2** bis(2,6-dichlorophenyl) disulfide (**138d**, 593 mg, 1.67 mmol, 3.00 equiv.) was reacted with **1** (1.00 mL, 555 µmol, 1.00 equiv.) in THF (2.00 mL). An inseparable mixture of **138d** (553 mg, 1.55 mmol) and **106d** (74 mg, 175 µmol, 32%) was obtained and the yield was determined by ^{1}H NMR spectroscopy.

R_f (n-pentane): 0.26; ^{1}H NMR (400 MHz, CDCl$_3$): δ = 7.41–7.38 (m, 4H, Ar-H), 7.21–7.17 (m, 2H, Ar-H), 2.07 (s, 6H, 3 × CH_2) ppm; ^{13}C NMR (100 MHz, CDCl$_3$): δ = 141.6 (C$_q$, 4 × C_{Ar}Cl), 134.3 (C$_q$, 2 × C_{Ar}S), 130.6 (+, 2 × CH_{Ar}), 128.7 (+, 4 × CH_{Ar}), 58.5 (+, 3 × CH_2), 42.2 (C$_q$, 2 × CCH$_2$) ppm.

1,3-Bis(4-methylphenylthio)bicyclo[1.1.1]pentane (106e)

According to **GP2** bis(4-methylphenyl) disulfide (**138e**, 386 mg, 1.57 mmol, 3.00 equiv.) was reacted with **1** (1.00 mL, 522 µmol, 1.00 equiv.) in THF (1.00 mL). The crude product was purified *via* column chromatography (n-pentane to n-pentane/Et$_2$O 50:1). The product was obtained as a colorless solid (156 mg, 499 µmol, 96%).

R_f (n-pentane/Et$_2$O 50:1): 0.51; m.p. 49–51 °C; ^{1}H NMR (400 MHz, CDCl$_3$): δ = 7.29–7.27 (m, 4H, Ar-H), 7.10 (d, ^3J = 7.9 Hz, 4H, Ar-H), 2.33 (s, 6H, 2 × CH_3), 1.97 (s, 6H, 3 × CH_2) ppm; ^{13}C NMR (100 MHz, CDCl$_3$): δ = 138.2 (C$_q$, 2 × C_{Ar}CH$_3$), 134.2 (+, 4 × CH_{Ar}), 129.8 (+, 4 × CH_{Ar}), 129.6 (C$_q$, 2 × C_{Ar}S), 57.3 (–, 3 × CH_2), 42.7 (C$_q$, 2 × CCH$_2$), 21.3 (+, 2 × CH_3) ppm; IR (ATR): $\tilde{v}$ = 2989, 2966, 2912, 2871, 1900, 1642, 1596, 1565, 1490, 1445, 1397, 1377, 1299, 1197, 1179, 1147, 1132, 1118, 1102, 1094, 1040, 1017, 929, 892, 809, 707, 633, 554, 513, 483, 415, 394 cm^{-1}; MS (EI, 60 °C): m/z (%) = 312 (8) [M]$^+$, 189 (100) [M–C$_7$H$_7$S]$^+$, 123 (45) [C$_7$H$_7$S]$^+$, 91 (18) [C$_7$H$_7$]$^+$; HRMS (EI, 60 °C): calc. for C$_{19}$H$_{20}$^{32}S$_2$ [M]$^+$ 312.1007, found 312.1006.

1,3-Bis(4-methoxyphenylthio)bicyclo[1.1.1]pentane (**106f**)

According to **GP2** bis(4-methoxyphenyl) disulfide (**138f**, 436 mg, 1.57 mmol, 3.00 equiv.) was reacted with **1** (1.00 mL, 522 µmol, 1.00 equiv.) in THF (1.00 mL). The crude product was purified *via* column chromatography (*n*-pentane/Et$_2$O 100:1). The product was obtained as a colorless solid (169 mg, 491 µmol, 94%).

R_f (*n*-pentane/Et$_2$O 100:1): 0.24; m.p. 65–67 °C; ^{1}H NMR (400 MHz, CDCl$_3$): δ = 7.34–7.30 (m, 4H, Ar-H), 6.84–6.80 (m, 4H, Ar-H), 3.79 (s, 6H, 2 × C*H*$_3$), 1.89 (s, 6H, 3 × C*H*$_2$) ppm; ^{13}C NMR (100 MHz, CDCl$_3$): δ = 159.9 (C$_q$, 2 × *C*$_{Ar}$O), 136.2 (+, 4 × *C*H$_{Ar}$), 123.7 (C$_q$, 2 × *C*$_{Ar}$S), 114.6 (+, 4 × *C*H$_{Ar}$), 56.9 (–, 3 × *C*H$_2$), 55.4 (+, 2 × *C*H$_3$), 43.0 (C$_q$, 2 × *C*CH$_2$) ppm; IR (ATR): $\tilde{v}$ = 2983, 2958, 2919, 2850, 2836, 1737, 1589, 1570, 1490, 1462, 1441, 1404, 1377, 1285, 1239, 1198, 1183, 1173, 1132, 1098, 1057, 1030, 1007, 922, 891, 875, 829, 813, 798, 758, 714, 663, 640, 628, 584, 555, 545, 524, 503, 487, 446, 405, 387, 381 cm^{-1}; MS (EI, 70 °C): *m/z* (%) = 344 (18) [M]$^+$, 205 (100) [M–C$_7$H$_7$OS]$^+$, 139 (69) [C$_7$H$_7$OS]$^+$; HRMS (EI, 70 °C): calc. for C$_{19}$H$_{20}$O$_2$^{32}S$_2$ [M]$^+$ 344.0906, found 344.0905.

1,3-Bis([1,1'-biphenyl]-2-thio)bicyclo[1.1.1]pentane (**106g**)

According to **GP2** di-2-biphenylyl disulfide (**132g**, 117 mg, 316 µmol, 3.00 equiv.) was reacted with **1** (200 µL, 105 µmol, 1.00 equiv.) in THF (1.00 mL). The crude product was purified *via* column chromatography (*n*-pentane to *n*-pentane/Et$_2$O 200:1). The product was obtained as a colorless solid (28 mg, 64.1 µmol, 61%).

R_f (*n*-pentane/Et$_2$O 200:1): 0.27; m.p. 108–110 °C; ^{1}H NMR (400 MHz, CDCl$_3$): δ = 7.49 (dt, ^3J = 7.2 Hz, ^4J = 1.2 Hz, 2H, Ar-H), 7.39–7.27 (m, 15H, Ar-H), 7.26–7.22 (m, 1H, Ar-H), 1.78 (s, 6H, 3 × C*H*$_2$) ppm; ^{13}C NMR (100 MHz, CDCl$_3$): δ = 145.1 (C$_q$, 2 × *C*$_{Ar}$), 141.1 (C$_q$, 2 × *C*$_{Ar}$), 134.5 (+, 2 × *C*H$_{Ar}$), 132.2 (C$_q$, 2 × *C*$_{Ar}$S), 130.7 (+, 2 × *C*H$_{Ar}$), 129.9 (+, 4 × *C*H$_{Ar}$), 127.8 (+, 2 × *C*H$_{Ar}$), 127.8 (+, 4 × *C*H$_{Ar}$), 127.7 (+, 2 × *C*H$_{Ar}$), 127.4 (+, 2 × *C*H$_{Ar}$), 57.6 (–, 3 × *C*H$_2$), 42.6 (C$_q$, 2 × *C*CH$_2$) ppm; IR (ATR): $\tilde{v}$ = 3053, 2993, 2962, 2910, 2870, 1584, 1494, 1459, 1443, 1424, 1198, 1157, 1123, 1069, 1038, 1007, 925, 870, 769, 751, 700, 680, 613, 552, 535, 501, 467, 435, 411, 380 cm^{-1}; MS (EI, 160 °C): *m/z* (%) = 436 (2) [M]$^+$, 251 (100) [M–C$_{12}$H$_9$S]$^+$, 185 (55) [C$_{12}$H$_9$S]$^+$, 184 (67) [C$_{12}$H$_9$S–H]$^+$; HRMS (EI, 160 °C): calc. for C$_{29}$H$_{24}$^{32}S$_2$ [M]$+$ 436.1321, found 436.1319.

3,3'-Bis(phenylthio)-1,1'-bi(bicyclo[1.1.1]pentane) (**143a**)

According to **GP3** diphenyl disulfide (**138a**, 43.7 mg, 200 µmol, 1.00 equiv.) was reacted with **1** (1.00 mL, 400 µmol, 2.00 equiv.) in THF (1.00 mL). The crude product was purified *via* preparative TLC (*n*-pentane). The products **106a** (18 mg, 63.3 µmol, 32%) and **143a** (7.0 mg, 20.0 µmol, 10%) were obtained as colorless solids.

R_f (*n*-pentane): 0.10; m.p. 133–135 °C; ^{1}H NMR (400 MHz, CDCl$_3$): δ = 7.42–7.39 (m, 4H, Ar-H), 7.31–7.27 (m, 6H, Ar-H), 1.73 (s, 12H, 3 × C*H*$_2$) ppm; ^{13}C NMR (100 MHz, CDCl$_3$): δ = 134.0 (C$_q$, 2 × *C*$_{Ar}$S), 133.8 (+, 4 × *C*H$_{Ar}$), 128.9 (+, 4 × *C*H$_{Ar}$), 127.7 (+, 2 × *C*H$_{Ar}$), 53.3 (−, 6 × *C*H$_2$), 41.6 (C$_q$, 2 × *C*CH$_2$), 40.3 (C$_q$, 2 × *C*CH$_2$) ppm; IR (ATR): $\tilde{\nu}$ = 2979, 2959, 2921, 2905, 2868, 1737, 1578, 1472, 1438, 1391, 1302, 1261, 1203, 1177, 1142, 1130, 1086, 1067, 1020, 999, 909, 887, 860, 802, 779, 744, 704, 688, 660, 615, 582, 550, 506, 477, 446, 421, 390 cm^{-1}; MS (EI, 100 °C): *m/z* (%) = 350 (4) [M]$^+$, 241 (32) [M–C$_6$H$_5$S]$^+$, 131 (100) [M–(C$_6$H$_5$S)$_2$–H]$^+$, 109 (15) [C$_6$H$_5$S]$^+$, 91 (53) [M–C$_6$H$_5$S–C$_9$H$_{10}$S]$^+$; HRMS (EI, 100 °C): calc. for C$_{22}$H$_{22}$^{32}S$_2$ [M]$^+$ 350.1161, found 350.1163.

3,3'-Bis(4-chlorophenylthio)-1,1'-bi(bicyclo[1.1.1]pentane) (**143b**)

According to **GP3** bis(4-chlorophenyl) disulfide (**138b**, 57.4 mg, 200 µmol, 1.00 equiv.) was reacted with **1** (1.00 mL, 400 µmol, 2.00 equiv.) in THF (1.00 mL). The crude product was purified *via* preparative TLC (*n*-pentane). The products **106b** (24 mg, 67.9 µmol, 34%) and **143b** (13 mg, 31.0 µmol, 15%) were obtained as colorless solids.

R_f (*n*-pentane): 0.30; m.p. 154–155 °C; ^{1}H NMR (400 MHz, CDCl$_3$): δ = 7.35–7.24 (m, 8H, Ar-H), 1.72 (s, 12H, 6 × C*H*$_2$) ppm; ^{13}C NMR (100 MHz, CDCl$_3$): δ = 135.1 (+, 4 × *C*H$_{Ar}$), 134.0 (C$_q$, 2 × *C*$_{Ar}$S), 132.4 (C$_q$, 2 × *C*$_{Ar}$Cl), 129.1 (+, 4 × *C*H$_{Ar}$), 53.3 (−, 6 × *C*H$_2$), 41.6 (C$_q$, 2 × *C*CH$_2$), 40.3 (C$_q$, 2 × *C*CH$_2$) ppm; IR (ATR): $\tilde{\nu}$ = 2982, 2959, 2919, 2904, 2868, 2851, 1907, 1735, 1655, 1571, 1472, 1441, 1385, 1259, 1207, 1201, 1173, 1142, 1116, 1091, 1011, 907, 887, 860, 822, 800, 745, 704, 557, 543, 506, 446, 390 cm^{-1}; MS (EI, 110 °C): *m/z* (%) = 422/420/418 (1/2/3) [M]$^+$, 275 (14) [M–C$_6$H$_4$ClS]$^+$, 143 (19) [C$_6$H$_4$ClS]$^+$, 131 (100) [M–(C$_6$H$_4$ClS)$_2$–H]$^+$, 91 (77) [M–C$_6$H$_4$ClS–C$_9$H$_9$ClS]$^+$; HRMS (EI, 110 °C): calc. for C$_{22}$H$_{20}$^{35}Cl$_2$^{32}S$_2$ [M]$^+$ 418.0382, found 418.0383.

3,3'-Bis(3,5-dichlorophenylthio)-1,1'-bi(bicyclo[1.1.1]pentane) (**143c**)

According to **GP3** bis(3,5-dichlorophenyl) disulfide (**138c**, 71.2 mg, 200 μmol, 1.00 equiv.) was reacted with **1** (1.00 mL, 400 μmol, 2.00 equiv.) in THF (1.00 mL). The crude product was purified *via* preparative TLC (*n*-pentane). The products **106c** (29 mg, 68.7 μmol, 34%) and **143c** (8.0 mg, 16.4 μmol, 8%) were obtained as colorless solids.

R_f (*n*-pentane): 0.49; m.p. 113–115 °C; ^{1}H NMR (400 MHz, CDCl$_3$): δ = 7.27 (d, ^4J = 1.8 Hz, 4H, Ar-H), 7.25 (t, ^4J = 1.8 Hz, 2H, Ar-H), 1.82 (s, 12H, 6 × CH_2) ppm; ^{13}C NMR (100 MHz, CDCl$_3$): δ = 137.7 (C$_q$, 2 × C_{Ar}S), 134.0 (C$_q$, 4 × C_{Ar}Cl), 130.9 (+, 4 × CH$_{Ar}$), 127.7 (+, 2 × CH$_{Ar}$), 53.6 (–, 6 × CH$_2$), 41.3 (C$_q$, 2 × CCH$_2$), 40.6 (C$_q$, 2 × CCH$_2$) ppm; IR (ATR): $\tilde{v}$ = 3072, 2986, 2959, 2921, 2908, 2874, 2853, 1689, 1605, 1551, 1463, 1445, 1418, 1401, 1378, 1364, 1288, 1261, 1211, 1177, 1140, 1129, 1095, 1052, 1023, 1006, 990, 890, 863, 853, 837, 793, 734, 663, 545, 528, 496, 459, 428, 402, 381 cm^{-1}; MS (EI, 140 °C): *m/z* (%) = 492/490/488/486 (1) [M]$^+$, 313/311/309 (2/8/11) [M–C$_6$H$_3$Cl$_2$S]$^+$, 131 (100) [M–(C$_6$H$_3$Cl$_2$S)$_2$–H]$^+$, 91 (50) [M–C$_6$H$_3$Cl$_2$S–C$_9$H$_8$Cl$_2$S]$^+$; HRMS (EI, 140 °C): calc. for C$_{22}$H$_{18}$^{35}Cl$_4$^{32}S$_2$ [M]$^+$ 485.9605, found 485.9604.

3,3'-Bis(4-methylphenylthio)-1,1'-bi(bicyclo[1.1.1]pentane) (**143e**)

According to **GP3** bis(4-methylphenyl) disulfide (**138e**, 49.3 mg, 200 μmol, 1.00 equiv.) was reacted with **1** (1.00 mL, 400 μmol, 2.00 equiv.) in THF (1.00 mL). The crude product was purified *via* preparative TLC (*n*-pentane). The products **106e** (22 mg, 70.4 μmol, 35%) and **143e** (9.0 mg, 23.8 μmol, 12%) were obtained as colorless solids.

R_f (*n*-pentane): 0.35; m.p. 142–144 °C; ^{1}H NMR (400 MHz, CDCl$_3$): δ = 7.29 (d, ^3J = 8.1 Hz, 4H, Ar-H), 7.09 (dd, ^3J = 8.4 Hz, ^4J = 0.8 Hz, 4H, Ar-H), 2.33 (s, 6H, 2 × CH_3), 1.68 (s, 12H, 6 × CH_2) ppm; ^{13}C NMR (100 MHz, CDCl$_3$): δ = 137.8 (C$_q$, 2 × C_{Ar}CH$_3$), 134.1 (+, 4 × CH$_{Ar}$), 130.2 (C$_q$, 2 × C_{Ar}S), 129.7 (+, 4 × CH$_{Ar}$), 53.2 (–, 6 × CH$_2$), 41.7 (C$_q$, 2 × CCH$_2$), 40.1 (C$_q$, 2 × CCH$_2$), 21.3 (+, 2 × CH$_3$) ppm; IR (ATR): $\tilde{v}$ = 2979, 2961, 2919, 2905, 2868, 1735, 1487, 1459, 1441, 1397, 1377, 1299, 1261, 1203, 1177, 1142, 1128, 1102, 1094, 1038, 1017, 970, 942, 909, 888, 860, 809, 778, 734, 707, 635, 550, 507, 479, 445, 405 cm^{-1}; MS (EI, 100 °C): *m/z* (%) = 378 (4) [M]$^+$, 255 (21) [M–C$_7$H$_7$S]$^+$, 189 (29) [M–C$_{12}$H$_{13}$S]$^+$, 131 (100) [M–(C$_7$H$_7$S)$_2$–H]$^+$, 91

(76) [M–C$_7$H$_7$S–C$_{10}$H$_{12}$S]$^+$; HRMS (EI, 100 °C): calc. for C$_{24}$H$_{26}$^{32}S$_2$ [M]$^+$ 378.1477, found 378.1476.

1,3-Bis(2-pyridylthio)bicyclo[1.1.1]pentane (**145**)

According to **GP2** 2,2'-dipydidyl disulfide (**144**, 252 mg, 1.14 mmol, 3.00 equiv.) was reacted with **1** (700 µL, 381 µmol, 1.00 equiv.) in THF (1.00 mL). The crude product was purified *via* column chromatography (*c*hex/EtOAc 20:1). The product was obtained as a pale brown oil (50 mg, 175 µmol, 46%).

R_f (*c*hex/EtOAc 20:1): 0.20; ^{1}H NMR (400 MHz, CDCl$_3$): δ = 8.45 (ddd, ^3J = 4.9 Hz, ^4J = 2.0 Hz, ^5J = 1.0 Hz, 2H, Ar-H), 7.50 (ddd, ^3J = 8.0 Hz, ^3J = 7.4 Hz, ^4J = 2.0 Hz, 2H, Ar-H), 7.23 (ddd, ^3J = 8.0 Hz, ^4J = 1.1 Hz, ^5J = 1.0 Hz, 2H, Ar-H), 7.03 (ddd, ^3J = 7.5 Hz, ^3J = 4.9 Hz, ^4J = 1.1 Hz, 2H, Ar-H), 2.55 (s, 6H, 3 × CH_2) ppm; ^{13}C NMR (100 MHz, CDCl$_3$): δ = 158.8 (C$_q$, 2 × C_{Ar}S), 149.8 (+, 2 × CH$_{Ar}$), 136.2 (+, 2 × CH$_{Ar}$), 124.2 (+, 2 × CH$_{Ar}$), 120.5 (+, 2 × CH$_{Ar}$), 58.5 (+, 3 × CH$_2$), 41.8 (C$_q$, 2 × CCH$_2$) ppm; IR (ATR): $\tilde{v}$ = 3293, 2978, 2917, 2873, 1687, 1613, 1574, 1555, 1543, 1446, 1412, 1383, 1373, 1350, 1326, 1298, 1278, 1261, 1197, 1166, 1122, 1081, 1031, 986, 962, 924, 874, 856, 816, 752, 722, 701, 669, 657, 611, 592, 552, 531, 510, 493, 480, 445, 397 cm^{-1}; MS (EI, 60 °C): *m/z* (%) = 286 (0.1) [M]$^+$, 176 (41) [M–C$_5$H$_4$NS]$^+$, 138 (90), 111 (59) [C$_5$H$_4$NS+H]$^+$, 71 (100); HRMS (EI, 60 °C): calc. for C$_{15}$H$_{14}$N$_2$^{32}S$_2$ [M]$^+$ 286.0598, found 286.0600.

1,3-Bis(benzylthio)bicyclo[1.1.1]pentane (**147**)

According to **GP2** dibenzyl disulfide (**146**, 323 mg, 1.31 mmol, 3.00 equiv.) was reacted with **1** (1.00 mL, 437 µmol, 1.00 equiv.) in THF (1.00 mL). The crude product was purified *via* HPLC (C18, H$_2$O/CH$_3$CN, 10:90 to 7:93). The product was obtained as a colorless solid (54 mg, 173 µmol, 40%).

R_f (*n*-pentane): 0.10; m.p. 63–65 °C; ^{1}H NMR (400 MHz, CDCl$_3$): δ = 7.34–7.20 (m, 10H, Ar-H), 3.73 (s, 4H, 2 × C$_6$H$_5$CH_2), 1.91 (s, 6H, 3 × CCH_2) ppm; ^{13}C NMR (100 MHz, CDCl$_3$): δ = 138.4 (C$_q$, 2 × C_{Ar}CH$_2$), 128.9 (+, 4 × CH$_{Ar}$), 128.6 (+, 4 × CH$_{Ar}$), 127.2 (+, 2 × CH$_{Ar}$), 57.0 (–, 3 × CCH_2), 41.3 (C$_q$, 2 × CCH$_2$), 36.2 (–, 2 × C$_6$H$_5$$CH_2$) ppm; IR (ATR): $\tilde{v}$ = 3027, 2986, 2961, 2907, 2871, 2844, 1493, 1452, 1435, 1239, 1201, 1126, 1068, 1027, 1000, 938, 914, 806, 776,

707, 694, 620, 572, 543, 499, 473, 415 cm^{-1}; MS (EI, 80 °C): m/z (%) = 221 (34) [M–C$_7$H$_7$]$^+$, 91 (100) [C$_7$H$_7$]$^+$; HRMS (EI, 80 °C): calc. for C$_{12}$H$_{13}{}^{32}$S$_2$ [M–C$_7$H$_7$]$^+$ 221.0459, found 221.0460.

1-(Phenylthio)-3-(4-methylphenylthio)bicyclo[1.1.1]pentane (**106j**)

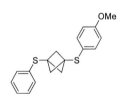

According to **GP2** a mixture of diphenyl disulfide (**138a**, 137 mg, 630 μmol, 1.50 equiv.) and bis(4-methylphenyl) disulfide (**138d**, 155 mg, 630 μmol, 1.50 equiv.) was reacted with **1** (1.00 mL, 420 μmol, 1.00 equiv.) in THF (1.00 mL). The crude product was purified *via* column chromatography (*n*-pentane). A mixture of **106a** (21.0 mg, 73.8 μmol, 18%), **106j** (56.0 mg, 188 μmol, 45%) and **106e** (31.0 mg, 99.2 μmol, 24%) was obtained and the yields were determined by ^{1}H NMR spectroscopy. An analytically pure sample of **106j** was obtained *via* HPLC (C18, H$_2$O/CH$_3$CN, 20:80 to 0:100). The product was obtained as a colorless solid.

R_f (*n*-pentane): 0.11; m.p. 47–49 °C; ^{1}H NMR (400 MHz, CDCl$_3$): δ = 7.44–7.40 (m, 2H, Ar-H), 7.33–7.28 (m, 5H, Ar-H), 7.11 (d, ^3J = 7.7 Hz, 2H, Ar-H), 2.34 (s, 3H, CH_3), 2.01 (s, 6H, 3 × CH_2) ppm; ^{13}C NMR (100 MHz, CDCl$_3$): δ = 138.2 (C$_q$, C_{Ar}CH$_3$), 134.2 (+, 2 × CH$_{Ar}$), 133.9 (+, 2 × CH$_{Ar}$), 133.4 (C$_q$, C_{Ar}S), 129.8 (+, 2 × CH$_{Ar}$), 129.6 (C$_q$, C_{Ar}S), 123.0 (+, 2 × CH$_{Ar}$), 128.0 (+, CH$_{Ar}$), 57.4 (–, 3 × CH$_2$), 42.8 (C$_q$, CCH$_2$), 42.6 (C$_q$, CCH$_2$), 21.3 (+, CH$_3$) ppm; IR (ATR): $\tilde{v}$ = 2989, 2962, 2908, 2867, 1489, 1473, 1436, 1198, 1176, 1126, 1101, 1092, 1062, 1026, 1016, 926, 809, 779, 744, 703, 690, 554, 511, 480, 421 cm^{-1}; MS (EI, 50 °C): m/z (%) = 298 (10) [M]$^+$, 189 (100) [M–C$_6$H$_5$S]$^+$, 175 (97) [M–C$_7$H$_7$S]$^+$, 91 (27) [C$_7$H$_7$]$^+$; HRMS (EI, 50 °C): calc. for C$_{18}$H$_{18}{}^{32}$S$_2$ [M]$^+$ 298.0850, found 298.0851.

1-(Phenylthio)-3-(4-methoxyphenylthio)bicyclo[1.1.1]pentane (**106k**)

According to **GP2** a mixture of diphenyl disulfide (**138a**, 143 mg, 655 μmol, 1.50 equiv.) and bis(4-methoxyphenyl) disulfide (**138f**, 182 mg, 655 μmol, 1.50 equiv.) was reacted with **1** (1.00 mL, 437 μmol, 1.00 equiv.) in THF (1.00 mL). The crude product was purified *via* column chromatography (*n*-pentane to *n*-pentane/Et$_2$O 50:1). Mixed fractions were collected and purified *via* preparative TLC (*n*-pentane/Et$_2$O 50:1). Products **106a** (22 mg, 77.3 μmol, 18%), **106k** (67 mg, 213 μmol, 49%) and **106f** (46 mg, 134 μmol, 31%) were obtained as colorless solids.

R_f (*n*-pentane/Et$_2$O 100:1): 0.33; m.p. 70–72 °C; ^{1}H NMR (500 MHz, CDCl$_3$): δ = 7.41–7.37 (m, 2H, Ar-H), 7.36–7.32 (m, 2H, Ar-H), 7.31–7.27 (m, 3H, Ar-H), 6.85–6.81 (m, 2H, Ar-H), 3.80 (s, 3H, C*H*$_3$), 1.96 (s, 6H, 3 × C*H*$_2$) ppm; ^{13}C NMR (125 MHz, CDCl$_3$): δ = 159.8 (C$_q$, *C*$_{Ar}$O), 136.2 (+, 2 × *C*H$_{Ar}$), 133.8 (+, 2 × *C*H$_{Ar}$), 133.3 (C$_q$, *C*$_{Ar}$S), 128.9 (+, 2 × *C*H$_{Ar}$), 127.9 (+, *C*H$_{Ar}$), 123.5 (C$_q$, *C*$_{Ar}$S), 114.5 (+, 2 × *C*H$_{Ar}$), 57.1 (–, 3 × *C*H$_2$), 55.3 (+, *C*H$_3$), 43.0 (C$_q$, *C*CH$_2$), 42.4 (C$_q$, *C*CH$_2$) ppm; IR (ATR): $\tilde{v}$ = 2989, 2956, 2907, 2867, 2834, 1588, 1570, 1489, 1458, 1438, 1404, 1282, 1237, 1197, 1170, 1126, 1095, 1064, 1027, 1006, 928, 833, 816, 798, 779, 747, 688, 640, 554, 527, 499, 486, 428, 388 cm^{-1}; MS (EI, 80 °C): *m/z* (%) = 314 (20) [M]$^+$, 205 (100) [M–C$_6$H$_5$S]$^+$, 175 (89) [M–C$_7$H$_7$OS]$^+$, 139 (76) [C$_7$H$_7$OS]$^+$; HRMS (EI, 80 °C): calc. for C$_{18}$H$_{18}$O^{32}S$_2$ [M]$^+$ 314.0799, found 314.0798.

Crystallographic information of the product can be found in chapter 7.4.

7.2.4 Modifications of bicyclo[1.1.1]pentylsulfides

Bicyclo[1.1.1]pentyl phenyl sulfoxide (**107a**)

 1-(Phenylthio)-bicyclo[1.1.1]pentane (**104a**, 500 mg, 2.84 mmol, 1.00 equiv.) was dissolved in CH$_2$Cl$_2$ (7.0 mL), *m*-chloroperbenzoic acid (≤ 77%, 734 mg, 3.28 mmol, 1.15 equiv.) was added in portions and the mixture was stirred for 5 min at room temperature. The precipitate was filtered off and the filtrate was washed with Na$_2$S$_2$O$_3$ solution (5.0 mL) and 1 M NaOH-solution (5.0 mL). The organic phase was dried over Na$_2$SO$_4$ and the solvent was removed under reduced pressure. The crude product was purified *via* column chromatography (*c*hex/EtOAc 5:1). The product was obtained as a colorless oil (365 mg, 1.90 mmol, 67%).

R_f (*c*hex/EtOAc, 5:1): 0.18; ^{1}H NMR (400 MHz, CDCl$_3$): δ = 7.53–7.47 (m, 5H, Ar-H), 2.81 (s, 1H, C*H*), 1.88 (s, 6H, 3 × C*H*$_2$) ppm; ^{13}C NMR (100 MHz, CDCl$_3$): δ = 141.7 (C$_q$, *C*$_{Ar}$), 130.9 (+, *C*H$_{Ar}$), 129.0 (+, 2 × *C*H$_{Ar}$), 124.3 (+, 2 × *C*H$_{Ar}$), 55.4 (C$_q$, *C*$_{Ar}$SC), 49.9 (–, 3 × *C*H$_2$), 27.8 (+, *C*H) ppm; IR (ATR): $\tilde{v}$ = 3462, 2970, 2917, 2880, 1581, 1476, 1443, 1303, 1201, 1129, 1084, 1067, 1036, 997, 883, 869, 777, 746, 692, 555, 511, 484 cm^{-1}; MS (EI, 20 °C): *m/z* (%) = 192 (2) [M]$^+$, 126 (100) [M–C$_5$H$_7$+H]$^+$, 125 (10) [M–C$_5$H$_7$]$^+$, 78 (41) [C$_6$H$_5$+H]$^+$, 77 (15) [C$_6$H$_5$]$^+$, 67 (61) [C$_5$H$_7$]$^+$; HRMS (EI, 20 °C): calc. for C$_{11}$H$_{12}$O^{32}S [M]$^+$ 192.0609; found 192.0610.

1-(2-Bromophenylthio)-bicyclo[1.1.1]pentane (**104s**)

According to **GP1** 2-bromothiophenol (**132s**, 450 µL, 708 mg, 3.74 mmol, 1.66 equiv.) was reacted with **1** (4.50 mL, 2.25 mmol, 1.00 equiv.) for 30 min. The product was obtained as a colorless oil (450 mg, 1.76 mmol, 78%).

^{1}H NMR (400 MHz, CDCl$_3$): δ = 7.60 (dd, ^3J = 8.0 Hz, ^4J = 1.4 Hz, 1H, Ar-H), 7.57 (dd, ^3J = 7.8 Hz, ^4J = 1.6 Hz, 1H, Ar-H), 7.29–7.23 (m, 1H, Ar-H), 7.13–7.07 (m, 1H, Ar-H), 2.76 (s, 1H, C*H*), 2.04 (s, 6H, 3 × C*H*$_2$) ppm; ^{13}C NMR (100 MHz, CDCl$_3$): δ = 136.2 (C$_q$, *C*$_{Ar}$S), 134.3 (+, *C*H$_{Ar}$), 133.3 (+, *C*H$_{Ar}$), 128.6 (+, *C*H$_{Ar}$), 127.7 (+, *C*H$_{Ar}$), 127.7 (C$_q$, *C*$_{Ar}$Br), 54.4 (–, 3 × *C*H$_2$), 45.5 (C$_q$, *C*$_{Ar}$S*C*), 29.5 (+, *C*H) ppm; IR (ATR): $\tilde{v}$ = 2979, 2910, 2874, 1570, 1557, 1445, 1425, 1247, 1204, 1160, 1126, 1119, 1103, 1071, 1018, 945, 924, 892, 864, 776, 745, 724, 647, 550, 518, 460, 424, 378 cm^{-1}; MS (EI, 20 °C): *m/z* (%) = 256/254 (33/33) [M]$^+$, 190/188 (57/56) [M–C$_5$H$_7$+H]$^+$, 134 (100) [M–C$_3$H$_5$–Br]$^+$, 109 (46) [M–C$_5$H$_7$–Br+H]$^+$, 108 (59) [M–C$_5$H$_7$–Br]$^+$, 67 (98) [C$_5$H$_7$]$^+$; HRMS (EI, 20 °C): calc. for C$_{11}$H$_{11}$^{79}Br^{32}S [M]$^+$ 253.9765; found 253.9766.

Bicyclo[1.1.1]pentyl 2-bromophenyl sulfone (**149s**)

1-(2-Bromophenylthio)-bicyclo[1.1.1]pentane (**104s**, 430 mg, 1.69 mmol, 1.00 equiv.) was dissolved in CH$_2$Cl$_2$ (4.3 mL) and stirred at 0 °C (ice bath). *m*-Chloroperbenzoic acid ($\leq$ 77%, 1.14 g, 5.07 mmol, 3.01 equiv.) was added in portions and after complete addition the mixture was stirred for 1 h at room temperature. The reaction mixture was diluted with CH$_2$Cl$_2$ and washed successively with sat. Na$_2$S$_2$O$_3$ solution (10 mL) and 1 M NaOH solution (10 mL). The organic phase was dried by the addition of Na$_2$SO$_4$ and the solvent was removed under reduced pressure. The product was obtained as a colorless solid (430 mg, 1.50 mmol, 89%).

R_f (*chex*/EtOAc 5:1): 0.30; m.p. 65–67 °C; ^{1}H NMR (400 MHz, CDCl$_3$): δ = 8.10 (dd, ^3J = 7.6, ^4J = 2.0 Hz, 1H, Ar-H), 7.76 (dd, ^3J = 7.7, ^4J = 1.5 Hz, 1H, Ar-H), 7.50 (td, ^3J = 7.6, ^4J = 1.5 Hz, 1H, Ar-H), 7.45 (td, ^3J = 7.5, ^4J = 2.0 Hz, 1H, Ar-H), 2.72 (s, 1H, C*H*), 2.21 (s, 6H, 3 × C*H*$_2$) ppm; ^{13}C NMR (100 MHz, CDCl$_3$): δ = 137.1 (C$_q$, *C*$_{Ar}$SO$_2$), 135.8 (+, *C*H$_{Ar}$), 134.7 (+, *C*H$_{Ar}$), 133.0 (+, *C*H$_{Ar}$), 128.0 (+, *C*H$_{Ar}$), 121.8 (C$_q$, *C*$_{Ar}$Br), 55.6 (C$_q$, CH$_2$*C*S), 51.7 (–, 3 × *C*H$_2$), 26.8 (+, *C*H) ppm; IR (ATR): $\tilde{v}$ = 3003, 2972, 2918, 2884, 1572, 1562, 1443, 1432, 1419, 1303, 1251, 1228, 1203, 1169, 1130, 1115, 1091, 1024, 955, 938, 897, 875, 779, 762, 730, 701, 649, 609, 565, 535, 490, 460, 436 cm^{-1}; MS (FAB): *m/z* (%) = 289/287 (20/19) [M+H]$^+$, 154 (52) [3-NBA+H], 137 (100) [3-NBA–OH+H]$^+$; HRMS (FAB): calc. for C$_{11}$H$_{12}$O$_2$^{79}Br^{32}S [M+H]$^+$: 286.9741; found 286.9740.

(2-(bicyclo[1.1.1]pentan-1-ylsulfonyl)phenyl)(phenyl)methanol (**153**)

Method A: Bicyclo[1.1.1]pentyl phenyl sulfone (**149a**, 42 mg, 201 µmol, 1.00 equiv.) was dissolved in THF (2.0 mL) under argon atmosphere. The solution was cooled to –78 °C and 1.7 M *tert*-butyllithium solution (125 µL, 13.6 mg, 213 µmol, 1.05 equiv.) in pentane was added dropwise. After complete addition, the reaction mixture was stirred for 1 h at –78 °C. Then, benzaldehyde (31 µL, 32.2 mg, 152 µmol, 1.51 equiv.) was added and the reaction mixture was allowed to warm to room temperature over 1 h. The reaction was quenched *via* careful addition of saturated NH$_4$Cl solution (2 mL). The reaction mixture was diluted with Et$_2$O (20 mL) and the organic layer was washed with water (10 mL). The organic layer was collected and dried by the addition of Na$_2$SO$_4$. The mixture was filtered and the solvent was evaporated under reduced pressure. The crude product was purified *via* column chromatography (*c*hex/EtOAc 5:1). The product was obtained as a colorless oil (16 mg, 50.9 µmol, 25%).

Method B: Bicyclo[1.1.1]pentyl 2-bromophenyl sulfone (**149s**, 57 mg, 199 µmol, 1.00 equiv.) was dissolved in THF (2.00 mL) under argon atmosphere and stirred at –78 °C. A 2.5 M *n*-butyllithium solution (84.0 µL, 13.5 mg, 210 µmol, 1.06 equiv.) was added dropwise. After complete addition, the reaction mixture was stirred for 1 h at –78 °C. Then, benzaldehyde (31 µL, 32.2 mg, 304 µmol, 1.53 equiv.) was added and the reaction mixture was allowed to warm to room temperature over 1 h. The reaction was quenched *via* careful addition of saturated NH$_4$Cl solution (2 mL). The reaction mixture was diluted with Et$_2$O (20 mL) and the organic layer was washed with water (10 mL). The organic layer was collected and dried by the addition of Na$_2$SO$_4$. The mixture was filtered and the solvent was evaporated under reduced pressure. The crude product was purified *via* column chromatography (*c*hex/EtOAc 5:1). The product was obtained as a colorless oil (32 mg, 102 µmol, 51%).

R_f (*c*hex/EtOAc 3:1): 0.29; ^{1}H NMR (400 MHz, CDCl$_3$): δ = 7.98 (dd, ^3J = 7.9 Hz, ^4J = 1.4 Hz, 1H, Ar-H), 7.54 (td, ^3J = 7.6 Hz, ^4J = 1.5 Hz, 1H, Ar-H), 7.46 (ddd, ^3J = 9.5 Hz, ^3J = 7.5 Hz, ^4J = 1.7 Hz, 3H, Ar-H), 7.40–7.34 (m, 2H, Ar-H), 7.31 (ddd, ^3J = 7.8 Hz, ^3J = 6.2 Hz, ^4J = 1.5 Hz, 2H, Ar-H), 6.59 (d, ^3J = 3.1 Hz, 1H, C*H*OH), 3.50 (d, ^3J = 4.4 Hz, 1H, CHO*H*), 2.77 (s, 1H, CH), 2.17 (s, 6H, 3 × C*H*$_2$) ppm; ^{13}C NMR (100 MHz, CDCl$_3$): δ = 145.2 (C$_q$, *C*$_{Ar}$), 141.7 (C$_q$, *C*$_{Ar}$), 135.3 (C$_q$, *C*$_{Ar}$), 134.3 (+, *C*H$_{Ar}$), 131.3 (+, *C*H$_{Ar}$), 130.5 (+, *C*H$_{Ar}$), 128.5 (+, 2 × *C*H$_{Ar}$), 128.5 (+, *C*H$_{Ar}$), 127.7 (+, *C*H$_{Ar}$), 126.9 (+, 2 × *C*H$_{Ar}$), 71.0 (+, *C*HOH), 55.9 (C$_q$, *C*H$_2$*C*S), 50.9 (–, 3 × *C*H$_2$), 26.7 (+, *C*H) ppm; IR (ATR): $\tilde{v}$ = 3483, 2997, 2973, 2919, 2884, 1494, 1449, 1298, 1279, 1207, 1193,

1167, 1133, 1105, 1077, 1058, 1034, 1021, 916, 875, 765, 739, 710, 649, 613, 565, 534, 456 cm^{-1}; MS (EI, 100 °C): m/z (%) = 314 (22) [M]$^+$, 230 (35) [M–C$_5$H$_7$–OH]$^+$, 229 (100) [M–C$_5$H$_7$–H$_2$O]$^+$; HRMS (EI, 100 °C): calc. for C$_{18}$H$_{18}$O$_3$^{32}S [M]$^+$ 314.0977; found 314.0976.

S-Bicyclo[1.1.1]pentyl-S-phenylsulfoximine (**108a**)

Method A: According to **GP4** 1-(phenylthio)-bicyclo[1.1.1]pentane (**104a**, 120 mg, 681 µmol, 1.00 equiv.) was reacted with ammonium carbonate (98 mg, 1.02 mmol, 1.50 equiv.) and (diacetoxyiodo)benzene (504 mg, 1.57 mmol, 2.30 equiv.) in MeOH (700 µL). The crude product was purified *via* column chromatography (*chex*/EtOAc 1:1). The product was obtained as a colorless solid (77 mg, 371 µmol, 55%).

Scale-up: According to **GP4** 1-(phenylthio)-bicyclo[1.1.1]pentane (**104a**, 1.20 g, 6.81 mmol, 1.00 equiv.) was reacted with ammonium carbonate (1.31 g, 13.6 mmol, 2.00 equiv.) and (diacetoxyiodo)benzene (6.58 g, 20.4 mmol, 3.00 equiv.) in MeOH (60.0 mL). The crude product was purified *via* column chromatography (*chex*/EtOAc 1:1). The product was obtained as a colorless solid (805 mg, 3.88 mmol, 57%).

Method B: Bicyclo[1.1.1]pentyl phenyl sulfoxide (**107a**, 140 mg, 728 µmol, 1.00 equiv.) was dissolved in MeOH (1.50 mL) and stirred at room temperature (open flask). First, ammonium carbamate (227 mg, 2.91 mmol, 4.00 equiv.) was added, followed by (diacetoxyiodo)benzene (634 mg, 2.18 mmol, 3.00 equiv.). The reaction mixture was stirred for 30 min. The solvent was removed under reduced pressure and the crude product was purified *via* column chromatography (*chex*/EtOAc 1:1). The product was obtained as a colorless solid (133 mg, 642 µmol, 88%).

R_f (*chex*/EtOAc 1:1): 0.16; m.p. 84–86 °C; ^{1}H NMR (400 MHz, CDCl$_3$): δ = 7.94–7.90 (m, 2H, Ar-H), 7.63–7.58 (m, 1H, Ar-H), 7.56–7.50 (m, 2H, Ar-H), 2.71 (s, 1H, CH), 2.08–2.00 (m, 6H, 3 × CH_2) ppm; ^{13}C NMR (100 MHz, CDCl$_3$): δ = 139.3 (C$_q$, C_{Ar}S), 133.1 (+, CH_{Ar}), 129.1 (+, 2 × CH_{Ar}), 129.1 (+, 2 × CH_{Ar}), 56.2 (C$_q$, CH$_2$CS), 50.3 (–, 3 × CH_2), 25.7 (+, CH) ppm; IR (ATR): $\tilde{v}$ = 3237, 2982, 2918, 2880, 1448, 1224, 1203, 1171, 1136, 1069, 1004, 989, 931, 874, 776, 756, 713, 690, 584, 548, 510, 455 cm^{-1}; MS (EI, 60 °C): m/z (%) = 207 (3) [M]$^+$, 141 (38) [M–C$_5$H$_7$+H]$^+$, 125 (92) [M–C$_5$H$_7$–NH]$^+$, 124 (72) [M–C$_5$H$_7$–O]$^+$, 93 (69), 77 (71) [C$_6$H$_5$]$^+$, 67 (100) [C$_5$H$_7$]$^+$; HRMS (EI, 60 °C): calc. for C$_{11}$H$_{13}$NO^{32}S [M+H]$^+$ 207.0718; found 207.0717.

The sulfoximine NH signal was not visible in ^{1}H NMR due to peak broadening. Crystallographic information of the product can be found in chapter 7.4.

S-Bicyclo[1.1.1]pentyl-*S*-4-chlorophenylsulfoximine (**108b**)

Method A: According to **GP4** 1-(4-chlorophenylthio)-bicyclo[1.1.1]pentane (**104b**, 83 mg, 394 µmol, 1.00 equiv.) was reacted with ammonium carbonate (76 mg, 791 µmol, 2.01 equiv.) and (diacetoxyiodo)benzene (381 mg, 1.18 mmol, 2.99 equiv.) in MeOH (5.0 mL). The crude product was purified *via* column chromatography (*c*hex/EtOAc 1:1). The product was obtained as a yellow oil (32 mg, 132 µmol, 34%).

Method B: Bicyclo[1.1.1]pentyl 4-chlorophenyl sulfoxide (**107b**, 29 mg, 128 µmol, 1.00 equiv.) was dissolved in MeOH (500 µL) and stirred at room temperature (open flask). First, ammonium carbonate (49 mg, 510 µmol, 3.98 equiv.) was added, followed by (diacetoxyiodo)benzene (124 mg, 384 µmol, 3.00 equiv.). The reaction mixture was stirred for 30 min. The solvent was removed under reduced pressure and the crude product was purified *via* column chromatography (*c*hex/EtOAc 1:1). The product was obtained as a yellow oil (30 mg, 124 µmol, 97%).

R_f (*c*hex/EtOAc 1:1): 0.17; ^{1}H NMR (400 MHz, CDCl$_3$): δ = 7.89–7.83 (m, 2H, Ar-H), 7.54–7.49 (m, 2H, Ar-H), 2.74 (s, 1H, C*H*), 2.06 (d, ^2J = 8.6 Hz, 3H, 3 × C*H*$_a$H$_b$), 2.05 (d, ^2J = 8.6 Hz, 3H, 3 × CH$_a$*H*$_b$) ppm; ^{13}C NMR (100 MHz, CDCl$_3$): δ = 139.3 (C$_q$, *C*$_{Ar}$S), 133.1 (+, *C*H$_{Ar}$), 129.1 (+, 2 × *C*H$_{Ar}$), 129.1 (+, 2 × *C*H$_{Ar}$), 56.2 (C$_q$, CH$_2$*C*S), 50.3 (−, 3 × *C*H$_2$), 25.7 (+, *C*H) ppm; IR (ATR): $\tilde{v}$ = 3263, 2994, 2918, 2884, 1579, 1474, 1391, 1227, 1206, 1087, 988, 878, 827, 748, 607, 559, 465 cm^{-1}; MS (EI, 70 °C): *m/z* (%) = 241 (1) [M]$^+$, 160 (24), 159 (51) [M–C$_5$H$_8$N]$^+$, 158 (33) [M–C$_5$H$_7$O]$^+$, 140 (21), 130 (5) [M–C$_6$H$_4$Cl]$^+$, 114 (10) [M–C$_6$H$_4$ClO]$^+$, 112(26), 111 (17) [M–C$_5$H$_8$NOS]$^+$, 82 (23), 67 (100) [C$_5$H$_7$]$^+$; HRMS (EI, 70 °C): calc. for C$_{11}$H$_{12}$ClNO^{32}S [M]$^+$ 241.0328; found 241.0329.

The sulfoximine NH signal was not visible in ^{1}H NMR due to peak broadening.

S-Bicyclo[1.1.1]pentyl-*S*-4-methylphenylsulfoximine (**108c**)

According to **GP4** 1-(4-methylphenylthio)-bicyclo[1.1.1]pentane (**104d**, 69 mg, 363 µmol, 1.00 equiv.) was reacted with ammonium carbonate (70 mg, 725 µmol, 1.98 equiv.) and (diacetoxyiodo)benzene (350 mg, 1.09 mmol, 3.00 equiv.) in MeOH (4.0 mL). The crude product was purified *via* column chromatography (*c*hex/EtOAc 1:2). The product was obtained as a colorless solid (48 mg, 217 µmol, 60%).

R_f (*c*hex/EtOAc 1:2): 0.25; m.p. 69–71 °C; m.p. 84–86 °C; ^{1}H NMR (400 MHz, CDCl$_3$): δ = 7.82–7.77 (m, 2H, Ar-H), 7.35–7.30 (m, 2H, Ar-H), 2.74 (s, 1H, C*H*), 2.44 (s, 3H, C*H*$_3$), 2.05 (d, ^2J =

9.1 Hz, 3H, $3 \times CH_aH_b$), 2.04 (d, ^2J = 9.1 Hz, 3H, $3 \times CH_aH_b$) ppm; ^{13}C NMR (100 MHz, CDCl$_3$): δ = 144.0 (C$_q$, C_{Ar}S), 136.2 (C$_q$, C_{Ar}CH$_3$), 129.8 (+, $2 \times CH_{Ar}$), 129.2 (+, $2 \times CH_{Ar}$), 56.3 (C$_q$, CH$_2$CS), 50.3 (–, $3 \times CH_2$), 25.7 (+, CH), 21.7 (+, CH$_3$) ppm; IR (ATR): $\tilde{v}$ = 3262, 2991, 2968, 2916, 2879, 1592, 1490, 1448, 1400, 1218, 1202, 1103, 982, 874, 814, 776, 706, 649, 575, 543, 501, 477 cm^{-1}; MS (EI, 50 °C): m/z (%) = 221 (3) [M]$^+$, 155 (22) [M–C$_5$H$_7$+H]$^+$, 139 (100) [M–C$_5$H$_7$–NH]$^+$, 91 (53) [C$_7$H$_7$]$^+$, 67 (61) [C$_5$H$_7$]$^+$; HRMS (EI, 50 °C): calc. for C$_{12}$H$_{15}$NO^{32}S [M+H]$^+$ 221.0874; found 221.0873.

The sulfoximine NH signal was not visible in ^{1}H NMR due to peak broadening.

S-Bicyclo[1.1.1]pentyl-*S*-4-methoxyphenylsulfoximine (**108d**)

According to **GP4** 1-(4-methoxyphenylthio)-bicyclo[1.1.1]pentane (**104f**, 89 mg, 431 μmol, 1.00 equiv.) was reacted with ammonium carbonate (83 mg, 863 μmol, 2.00 equiv.) and (diacetoxyiodo)benzene (415 mg, 1.29 mmol, 2.99 equiv.) in MeOH (5.0 mL). The crude product was purified *via* column chromatography (*c*hex/EtOAc 1:2). The product was obtained as a yellow oil (85 mg, 358 μmol, 83%).

R_f (*c*hex/EtOAc 1:2): 0.11; ^{1}H NMR (400 MHz, CDCl$_3$): δ = 7.86–7.81 (m, 2H, Ar-H), 7.01–6.95 (m, 2H, Ar-H), 3.87 (s, 3H, OCH_3), 2.69 (s, 1H, CH), 2.03 (d, ^2J = 9.2 Hz, 3H, $3 \times CH_aH_b$), 2.02 (d, ^2J = 9.2 Hz, 3H, $3 \times CH_aH_b$) ppm; ^{13}C NMR (100 MHz, CDCl$_3$): δ = 163.4 (C$_q$, C_{Ar}OCH$_3$), 131.2 (+, $2 \times CH_{Ar}$), 130.6 (C$_q$, C_{Ar}S), 114.3 (+, $2 \times CH_{Ar}$), 56.3 (C$_q$, CH$_2$CS), 55.7 (+, OCH$_3$), 50.3 (–, $3 \times CH_2$), 25.6 (+, CH) ppm; IR (ATR): $\tilde{v}$ = 3258, 2974, 2918, 2883, 2840, 2050, 1593, 1577, 1494, 1460, 1411, 1309, 1257, 1219, 1104, 1020, 983, 877, 835, 800, 656, 628, 580, 550, 523 cm^{-1}; MS (EI, 60 °C): m/z (%) = 237 (6) [M]$^+$, 170 (9) [M–C$_5$H$_7$]$^+$, 155 (100) [M–C$_5$H$_7$–NH]$^+$, 154 (36) [M–C$_5$H$_7$–O]$^+$, 107 (4) [C$_7$H$_7$O]$^+$, 67 (19) [C$_5$H$_7$]$^+$; HRMS (EI, 60 °C): calc. for C$_{12}$H$_{15}$NO$_2$^{32}S [M+H]$^+$: 237.0824; found 237.0825.

The sulfoximine NH signal was not visible in ^{1}H NMR due to peak broadening.

S-Bicyclo[1.1.1]pentyl-*S*-*n*-butylsulfoximine (**108e**)

According to **GP4** 1-(*n*-butylthio)-bicyclo[1.1.1]pentane (**105c**, 33 mg, 211 μmol, 1.00 equiv.) was reacted with ammonium carbonate (41 mg, 427 μmol, 2.02 equiv.) and (diacetoxyiodo)benzene (204 mg, 633 μmol, 3.00 equiv.) in MeOH (3.0 mL). The crude product was purified *via* column chromatography (*c*hex/EtOAc 1:4). The product was obtained as a colorless oil (17 mg, 90.7 μmol, 43%).

R_f (chex/EtOAc 1:4): 0.18; ^{1}H NMR (400 MHz, CDCl$_3$): δ = 2.56–2.49 (m, 2H, SCH_2), 2.78 (s, 1H, CH), 2.25–2.18 (m, 6H, 3 × SCCH_2), 1.87–1.78 (m, 2H, SCH$_2$CH_2), 1.45 (qt, ${}^3J$ = 7.4 Hz, ${}^3J$ = 7.4 Hz, 2H, CH_2CH$_3$), 0.95 (t, ${}^3J$ = 7.4 Hz, 3H, CH$_3$) ppm; ^{13}C NMR (100 MHz, CDCl$_3$): δ = 55.9 (C$_q$, CH$_2$CS), 51.2 (–, SCH$_2$), 50.8 (–, 3 × CH$_2$), 25.8 (+, CH), 23.9 (–, SCH$_2$CH$_2$), 22.0 (–, CH$_2$CH$_3$), 13.8 (+, CH$_3$) ppm; IR (ATR): $\tilde{v}$ = 3261, 2961, 2919, 2875, 1735, 1452, 1408, 1379, 1207, 1139, 1096, 978, 875, 779, 597, 527, 493 cm^{-1}; MS (EI, 20 °C): m/z (%) = 188 (1) [M+H]$^+$, 187 (9) [M]$^+$, 89 (6) [M–C$_5$H$_7$–NH–O]$^+$, 67 (100) [C$_5$H$_7$]$^+$, 57 (14) [C$_4$H$_9$]$^+$; HRMS (EI, 20 °C): calc. for C$_9$H$_{17}$NO^{32}S [M]$^+$ 187.1031; found 187.1030.

The sulfoximine NH signal was not visible in ^{1}H NMR due to peak broadening.

S-Benzyl-S-bicyclo[1.1.1]pentylsulfoximine (**108f**)

According to **GP4** 1-(benzylthio)-bicyclo[1.1.1]pentane (**105f**, 67 mg, 352 µmol, 1.00 equiv.) was reacted with ammonium carbonate (68 mg, 704 µmol, 2.00 equiv.) and (diacetoxyiodo)benzene (304 mg, 1.06 mmol, 3.01 equiv.) in MeOH (3.0 mL). The crude product was purified *via* column chromatography (chex/EtOAc 1:2). The product was obtained as a colorless solid (32 mg, 145 µmol, 41%).

R_f (chex/EtOAc 1:2): 0.12; m.p. 128–130 °C; ^{1}H NMR (400 MHz, CDCl$_3$): δ = 7.51–7.30 (m, 5H, Ar-H), 4.32 (d, ${}^2J$ = 12.7 Hz, 1H, SCH$_a$H$_b$), 4.11 (d, ${}^2J$ = 12.7 Hz, 1H, SCH$_a$H$_b$), 2.77 (s, 1H, CH), 2.51 (bs, 1H, NH), 2.23 (d, ${}^2J$ = 9.2 Hz, 3H, 3 × SCCH$_a$H$_b$), 2.22 (d, ${}^2J$ = 9.2Hz, 3H, 3 × SCCH$_a$H$_b$) ppm; ^{13}C NMR (100 MHz, CDCl$_3$): δ = 131.3 (+, 2 × CH$_{Ar}$), 129.0 (+, CH$_{Ar}$), 128.9 (+, 2 × CH$_{Ar}$), 127.5 (C$_q$, $C$$_{Ar}$), 58.8 (–, S$CH_2$), 55.4 (C$_q$, S$CCH_2$), 51.0 (–, 3 × CH$_2$), 26.2 (+, CH) ppm; IR (ATR): $\tilde{v}$ = 3314, 2964, 2877, 1602, 1493, 1455, 1211, 1137, 1078, 1029, 956, 876, 765, 698, 618, 546, 518, 494 cm^{-1}; MS (EI, 80 °C): m/z (%) = 221 (2) [M]$^+$, 143 (10) [M–C$_6$H$_5$–H]$^+$, 130 (3) [M–C$_7$H$_7$]$^+$, 91 (100) [C$_7$H$_7$]$^+$, 67 (60) [C$_5$H$_7$]$^+$; HRMS (EI, 80 °C): calc. for C$_{12}$H$_{15}$NO^{32}S [M]$^+$ 221.0874; found 221.0873.

Crystallographic information of the product can be found in chapter 7.4.

Methyl 3-(bicyclo[1.1.1]pentane-1-sulfonimidoyl)propionat (**108g**)

According to **GP4** methyl 3-(1-bicyclo[1.1.1]pentylthio) propionate (**105k**, 53 mg, 285 µmol, 1.00 equiv.) was reacted with ammonium carbonate (55 mg, 569 µmol, 2.00 equiv.) and (diacetoxyiodo)benzene (275 mg, 854 µmol, 3.00 equiv.) in MeOH (3.0 mL). The crude product was purified *via* column chromatography (chex/EtOAc 1:4). The product was obtained as a colorless oil (17 mg, 78.2 µmol, 27%).

R_f (chex/EtOAc 1:4): 0.11; ^{1}H NMR (400 MHz, CDCl$_3$): δ = 3.72 (s, 3H, OCH$_3$), 3.32–3.25 (m, 2H, SCH$_2$CH$_2$), 2.90–2.84 (m, 2H, SCH$_2$CH$_2$), 2.79 (s, 1H, CH), 2.26–2.19 (m, 6H, 3 × SCCH$_2$) ppm; ^{13}C NMR (100 MHz, CDCl$_3$): δ = 171.4 (C$_q$, CO$_2$CH$_3$), 55.9 (C$_q$, SCCH$_2$), 52.5 (+, CH$_3$), 50.7 (–, 3 × SCCH$_2$), 46.6 (SCH$_2$CH$_2$), 27.0 (SCH$_2$CH$_2$), 25.8 (+, CH) ppm; IR (ATR): $\tilde{v}$ = 3273, 2975, 2920, 2885, 1735, 1438, 1363, 1208, 1141, 1011, 876, 780, 696, 677, 546, 494 cm^{-1}; MS (ESI+): m/z (%) = 239 (1) [M+Na]$^+$, 218 (100) [M+H]$^+$; HRMS (ESI+): calc. for C$_9$H$_{15}$NO$_3$^{32}S [M+H]$^+$ 218.0845; found 218.0835.

The sulfoximine NH signal was not visible in ^{1}H NMR due to peak broadening.

N-Phenyl-S-bicyclo[1.1.1]pentyl-S-phenylsulfoximine (**159a**)

S-Bicyclo[1.1.1]pentyl-S-phenylsulfoximine (**108a**, 50 mg, 241 µmol, 1.00 equiv.), cesium carbonate (157 mg, 482 µmol, 2.00 equiv.) and copper(I) acetate (8.9 mg, 72.6 µmol, 0.300 equiv.) were mixed in a dry vial under argon atmosphere. DMF (500 µL) was added, followed by iodobenzene (41 µL, 74.7 mg, 366 µmol, 1.52 equiv.) and the reaction mixture was stirred at 100 °C for 18 h. After cooling to room temperature, the mixture was diluted with 10 mL of CH$_2$Cl$_2$, filtered through a short silica pad and concentrated under reduced pressure. The crude product was purified via column chromatography (chex/EtOAc 10:1). The product was obtained as a colorless solid (37 mg, 131 µmol, 54%).

R_f (chex/EtOAc 10:1): 0.15; m.p. 93–95 °C; ^{1}H NMR (400 MHz, CDCl$_3$): δ = 7.90–7.86 (m, 2H, Ar-H), 7.60–7.54 (m, 1H, Ar-H), 7.53–7.47 (m, 2H, Ar-H), 7.14–7.09 (m, 2H, Ar-H), 7.08–7.04 (m, 2H, Ar-H), 6.90–6.78 (m, 1H, Ar-H), 2.71 (s, 1H, CH), 2.08 (s, 6H, 3 × CH$_2$) ppm; ^{13}C NMR (100 MHz, CDCl$_3$): δ = 145.4 (C$_q$, C$_{Ar}$), 136.4 (C$_q$, C$_{Ar}$), 133.1 (+, CH$_{Ar}$), 129.9 (+, 2 × CH$_{Ar}$), 129.3 (+, 2 × CH$_{Ar}$), 129.0 (+, 2 × CH$_{Ar}$), 123.7 (+, 2 × CH$_{Ar}$), 121.5 (+, CH$_{Ar}$), 56.3 (C$_q$, SCCH$_2$), 50.9 (–, 3 × CH$_2$), 26.3 (+, CH) ppm; IR (ATR): $\tilde{v}$ = 2999, 2917, 2883, 1594, 1570, 1482, 1443, 1300, 1266, 1198, 1191, 1174, 1156, 1142, 1091, 1068, 1034, 1014, 994, 948, 928, 891, 878, 779, 765, 755, 701, 686, 637, 612, 555, 528, 513, 465, 424 cm^{-1}; MS (EI, 60 °C): m/z (%) = 284 (5) [M+H]$^+$, 283 (29) [M]$^+$, 217 (31) [C$_{12}$H$_{10}$NOS+H]$^+$, 200 (60) [C$_{12}$H$_{10}$NS]$^+$, 125 (83) [C$_6$H$_5$OS]$^+$, 92 (100) [C$_6$H$_5$N+H], 91 (11) [C$_6$H$_5$N]$^+$, 77 (56) [C$_5$H$_5$]$^+$, 67 (54) [C$_5$H$_7$]$^+$; HRMS (EI, 60 °C): calc. for C$_{17}$H$_{17}$NO^{32}S [M]$^+$ 283.1031; found 283.1030.

7.2.5   Sodium bicyclo[1.1.1]pentanesulfinate

Methyl 3-(1-bicyclo[1.1.1]pentylthio) propionate (**105k**) – scale-up

In a flame-dried round-bottomed flask that has been purged with argon 1,1-dibromo-2,2-bis(chloromethyl)cyclopropane (**44**, 31.0 g, 94.0 mmol, 1.00 equiv.) was dissolved in Et$_2$O (130 mL) and cooled to –40 °C. A 1.9 M PhLi solution (100 mL, 190 mmol, 2.02 equiv.) in Bu$_2$O was added dropwise under vigorous stirring. After complete addition the mixture was warmed to 0 °C and stirred at this temperature for further 2 h. The reaction flask was attached to an argon purged rotavap with dry ice condenser. The receiving flask, containing a magnetic stir bar, was cooled to –78 °C and the product was distilled together with Et$_2$O. The water bath was set to 20 °C and the pressure was reduced from 500 mbar to 20 mbar slowly. A solution of **1** in Et$_2$O was obtained and used directly in the next step.

Methyl 3-mercaptopropionate (**132k**, 13.5 mL, 14.7 g, 122 mmol, 1.30 equiv.) was added to the solution of **1** under argon atmosphere and the mixture was stirred at room temperature for 30 min. The reaction mixture was poured into a separation funnel and the organic layer was washed successively with 150 mL of a 1 M NaOH-solution. The organic layer was dried by the addition of Na$_2$SO$_4$. The mixture was filtered through a glass funnel and the solvent was evaporated under reduced pressure. The product was obtained as a colorless oil (13.8 g, 74.3 mmol, 79% from **44**).

Analytical data for **105k** can be found in chapter 7.2.2.

Methyl 3-(1-bicyclo[1.1.1]pentylsulfonyl) propionate (**149k**)

Method A: The sulfide **105k** (13.8 g, 74.3 mmol, 1.00 equiv.) was dissolved in CH$_2$Cl$_2$ (140 mL) and cooled to 0 °C in an ice bath. *m*-Chloroperbenzoic acid ($\leq$ 77%, 43.6 g, 194 mmol, 2.62 equiv.) was added in portions and the mixture was stirred for 1 h at room temperature after complete addition. The reaction mixture was poured into 150 mL of sat. Na$_2$S$_2$O$_3$-solution. The precipitated *m*-chlorobenzoic acid was filtered off and washed with 50 mL of CH$_2$Cl$_2$. The filtrate was poured into a separation funnel and the phases were separated. The organic layer was washed successively with 250 mL of a 1 M NaOH-solution. The organic layers were collected and were dried by the addition of Na$_2$SO$_4$. The mixture was filtered through a glass funnel and the solvent was evaporated under reduced pressure. The product was obtained as a colorless oil (13.3 g, 60.9 mmol, 82%).

Method B: To a solution of the sulfide **105k** (4.39 g, 23.6 mmol, 1.00 equiv.) in 1,4-dioxane/water 1:1 (170 mL) potassium peroxymonosulfate (OXONE®) (29.0 g, 47.1 mmol, 2.00 equiv.) was added over 5 min. The reaction mixture was stirred at room temperature for 18 h. The solid was removed by filtration and washed with 50 mL of 1,4-dioxane. The filtrate was concentrated under reduced pressure to remove the organic phase. The aqueous phase was partitioned between 150 mL of CH_2Cl_2 and 100 mL of sat. $NaHCO_3$-solution. The organic layer was collected and was dried by the addition of Na_2SO_4. The mixture was filtered through a glass funnel and the solvent was evaporated under reduced pressure. The product was obtained as a colorless oil (3.69 g, 16.9 mmol, 72%).

^{1}H NMR (400 MHz, CDCl$_3$): δ = 3.73 (s, 3H, OCH_3), 3.28–3.21 (m, 2H, SCH_2CH$_2$), 2.88–2.82 (m, 2H, SCH$_2$CH_2), 2.79 (s, 1H, CH), 2.27 (s, 6H, 3 × CH_2) ppm; ^{13}C NMR (100 MHz, CDCl$_3$): δ = 171.1 (C$_q$, CO$_2$CH$_3$), 54.1 (C$_q$, CSCH$_2$), 52.6 (+, OCH$_3$), 51.0 (–, 3 × CCH$_2$CH), 45.1 (–, SCH$_2$CH$_2$), 26.9 (+, CH), 26.2 (–, SCH$_2$$CH_2$) ppm; IR (ATR): $\tilde{\nu}$ = 2996, 2976, 2956, 2922, 2887, 1737, 1438, 1364, 1302, 1251, 1210, 1173, 1140, 1106, 1057, 1018, 980, 941, 898, 877, 830, 782, 718, 704, 625, 605, 594, 545, 513, 477, 448, 397, 378 cm^{-1}; MS (EI, 70 °C): m/z (%) = 219 (0.3) [M+H]$^+$, 135 (17) [M–C$_5$H$_7$O]$^+$, 88 (26) [M–C$_5$H$_7$SO$_2$+H]$^+$, 67 (100) [C$_5$H$_7$]$^+$; HRMS (EI, 70 °C): calc. for C$_9$H$_{15}$O$_4$^{32}S [M+H]$^+$: 219.0691; found 219.0690.

Sodium bicyclo[1.1.1]pentanesulfinate (BCP-SO$_2$Na) (**169**)

The sulfone **149k** (13.3 g, 60.9 mmol, 1.00 equiv.) was dissolved in THF (130 mL) and stirred at room temperature. A 5.4 M sodium methanolate solution (3.29 g, 11.3 mL, 60.9 mmol, 1.00 equiv.) in MeOH was added to the solution and a pale yellow solid precipitated. After complete addition the reaction mixture was stirred at room temperature for 20 min. The solvent and the formed acrylate were evaporated under reduced pressure. The product was obtained as a pale yellow solid (9.38 g, 60.8 mmol, quant.).

m.p. > 300 °C; ^{1}H NMR (400 MHz, D$_2$O): δ = 2.68 (s, 1H, CH), 1.88 (s, 6H, 3 × CH_2) ppm; ^{13}C NMR (100 MHz, D$_2$O): δ = 57.2 (–, 3 × CCH$_2$CH), 47.2 (C$_q$, CSO$_2$Na), 25.8 (+, CH) ppm; IR (ATR): $\tilde{\nu}$ = 2978, 2959, 2904, 2873, 1205, 1190, 1017, 990, 933, 898, 860, 779, 664, 584, 524, 477, 397 cm^{-1}; MS (ESI–): m/z (%) = 131 (100) [M–Na]$^-$; MS (ESI+): m/z (%) = 133 (100) [M+2H]$^+$; HRMS (ESI+): calc. for C$_5$H$_9$O$_2$^{32}S [M+2H]$^+$: 133.0318; found 133.0314.

Bicyclo[1.1.1]pentyl 4-nitrophenyl sulfone (**149b**)

BCP-SO$_2$Na (**169**, 142 mg, 921 μmol, 1.30 equiv.) and 1-fluoro-4-nitrobenzene (**170a**, 75 μL, 100 mg, 709 μmol, 1.00 equiv.) were mixed in a dry vial under argon atmosphere. DMF (1.0 mL) was added and the mixture was stirred at 80 °C for 3 d. The reaction mixture was cooled down and purified *via* column chromatography (*chex*/EtOAc 5:1). The product was obtained as a white solid (172 mg, 679 μmol, 96%).

R_f (*chex*/EtOAc 5:1): 0.22; m.p. 183–185 °C; ^{1}H NMR (400 MHz, CDCl$_3$): δ = 8.44–8.39 (m, 2H, Ar-H), 8.08–8.04 (m, 2H, Ar-H), 2.79 (s, 1H, C*H*), 2.12 (s, 6H, 3 × C*H*$_2$) ppm; ^{13}C NMR (100 MHz, CDCl$_3$): δ = 151.0 (C$_q$, *C*$_{Ar}$NO$_2$), 142.9 (C$_q$, *C*$_{Ar}$SO$_2$), 130.2 (+, 2 × *C*H$_{Ar}$), 124.4 (+, 2 × *C*H$_{Ar}$), 55.0 (C$_q$, CH$_2$*C*S), 50.8 (–, 3 × *C*CH$_2$CH), 27.3 (+, *C*H) ppm; IR (ATR): $\tilde{v}$ = 3003, 2919, 1524, 1349, 1296, 1285, 1210, 1194, 1170, 1126, 1078, 1014, 941, 881, 854, 778, 747, 732, 684, 619, 575, 548, 452, 407 cm^{-1}; MS (EI, 130 °C): *m/z* (%) = 254 (0.1) [M+H]$^+$, 253 (0.1) [M]$^+$, 170 (3) [M–C$_5$H$_7$–O]$^+$, 67 (100) [C$_5$H$_7$]$^+$; HRMS (EI, 130 °C): calc. for C$_{11}$H$_{11}$O$_4$N^{32}S [M]$^+$: 253.0409; found 253.0407.

Crystallographic information of the product can be found in chapter 7.4.

Bicyclo[1.1.1]pentyl 2-nitrophenyl sulfone (**149c**)

BCP-SO$_2$Na (**169**, 142 mg, 921 μmol, 1.30 equiv.) and 1-fluoro-2-nitrobenzene (**170b**, 100 mg, 709 μmol, 1.00 equiv.) were mixed in a dry vial under argon atmosphere. DMF (1.0 mL) was added and the mixture was stirred at 80 °C for 16 h. The reaction mixture was cooled down and purified *via* column chromatography (*chex*/EtOAc 5:1). The product was obtained as a colorless solid (176 mg, 695 μmol, 98%).

R_f (*chex*/EtOAc 5:1): 0.15; m.p. 89–91 °C; ^{1}H NMR (400 MHz, CDCl$_3$): δ = 8.08–8.04 (m, 1H, Ar-H), 7.82–7.72 (m, 3H, Ar-H), 2.76 (s, 1H, C*H*), 2.29 (s, 6H, 3 × C*H*$_2$) ppm; ^{13}C NMR (100 MHz, CDCl$_3$): δ = 149.8 (C$_q$, *C*$_{Ar}$NO$_2$), 134.9 (+, *C*H$_{Ar}$), 132.6 (+, *C*H$_{Ar}$), 132.2 (+, *C*H$_{Ar}$), 131.3 (C$_q$, *C*$_{Ar}$SO$_2$), 124.5 (+, *C*H$_{Ar}$), 56.5 (C$_q$, CH$_2$*C*S), 52.0 (–, 3 × *C*CH$_2$CH), 26.4 (+, *C*H) ppm; IR (ATR): $\tilde{v}$ = 3094, 3003, 2972, 2918, 2885, 1589, 1536, 1438, 1371, 1309, 1207, 1170, 1137, 1103, 1055, 969, 945, 895, 877, 851, 773, 745, 721, 653, 611, 565, 533, 455, 384 cm^{-1}; MS (FAB): *m/z* (%) = 254 (0.1) [M+H]$^+$, 253 (0.1) [M]$^+$, 170 (3) [M–C$_5$H$_7$–O]$^+$, 67 (100) [C$_5$H$_7$]$^+$; HRMS (FAB): calc. for C$_{11}$H$_{11}$O$_4$N^{32}S [M]$^+$: 253.0409; found 253.0407.

Crystallographic information of the product can be found in chapter 7.4.

Bicyclo[1.1.1]pentyl 2-nitro-4-hydroxyphenyl sulfone (**149d**)

BCP-SO$_2$Na (**169**, 128 mg, 828 µmol, 1.30 equiv.) and 3-fluoro-4-nitrophenol (**170c**, 100 mg, 637 µmol, 1.00 equiv.) were mixed in a dry vial under argon atmosphere. DMF (1.0 mL) was added and the mixture was stirred at 100 °C for 60 h. The reaction mixture was cooled down and purified *via* column chromatography (chex/EtOAc 2:1). The product was obtained as a colorless solid (128 mg, 475 µmol, 75%).

R_f (chex/EtOAc 2:1): 0.09; m.p. 133–135 °C; ^{1}H NMR (400 MHz, CDCl$_3$): δ = 7.89 (d, ^3J = 8.8 Hz, 1H, Ar-H), 7.65 (d, ^4J = 2.7 Hz, 1H, Ar-H), 7.19 (dd, ^3J = 8.8 Hz, ^4J = 2.7 Hz, 1H, Ar-H), 2.79 (s, 1H, C*H*), 2.34 (s, 6H, 3 × C*H$_2$*) ppm; ^{13}C NMR (100 MHz, CDCl$_3$): δ = 176.1 (C$_q$, C_{Ar}OH), 160.0 (C$_q$, C_{Ar}NO$_2$), 133.8 (C$_q$, C_{Ar}SO$_2$), 128.1 (+, CH_{Ar}), 121.0 (+, CH_{Ar}), 119.0 (+, CH_{Ar}), 57.2 (C$_q$, CH$_2$*C*S), 52.6 (–, 3 × *C*CH$_2$CH), 26.2 (+, *C*H) ppm; IR (ATR): $\tilde{v}$ = 3268, 3009, 2968, 2917, 1598, 1581, 1544, 1486, 1432, 1356, 1299, 1249, 1228, 1205, 1156, 1139, 1096, 1041, 939, 912, 875, 847, 839, 773, 745, 694, 629, 592, 560, 534, 517, 473, 435 cm^{-1}; MS (FAB): *m/z* (%) = 270 (8) [M+H]$^+$, 186 (12) [M–C$_5$H$_7$–O]$^+$, 154 (100) [3-NBA+H], 137 (75) [3-NBA–OH+H]$^+$; HRMS (FAB): calc. for C$_{11}$H$_{12}$O$_5$N^{32}S [M+H]$^+$: 270.0436; found 270.0435.

The OH signal was not visible in ^{1}H NMR due to peak broadening. Crystallographic information of the product can be found in chapter 7.4.

2-(Bicyclo[1.1.1]pentan-1-ylsulfonyl)quinoline (**149e**)

BCP-SO$_2$Na (**169**, 123 mg, 795 µmol, 1.30 equiv.), 2-chloroquinoline (**170d**, 100 mg, 611 µmol, 1.00 equiv.) and potassium carbonate (127 mg, 917 µmol, 1.50 equiv.) were mixed in a dry vial under argon atmosphere. DMF (1.0 mL) was added and the mixture was stirred at 120 °C for 16 h. The reaction mixture was cooled down and purified *via* column chromatography (chex/EtOAc 10:1). The product was obtained as a white solid (46 mg, 177 µmol, 29%).

R_f (chex/EtOAc 5:1): 0.16; m.p. 101–103 °C; ^{1}H NMR (400 MHz, CDCl$_3$): δ = 8.42 (dd, ^3J = 8.5 Hz, ^4J = 0.9 Hz, 1H, Ar-H), 8.28 (dt, ^3J = 8.6 Hz, ^4J = 1.0 Hz, 1H, Ar-H), 8.09 (d, ^3J = 8.5 Hz, 1H, Ar-H), 7.94 (dd, ^3J = 8.2 Hz, ^4J = 1.4 Hz, 1H, Ar-H), 7.85 (ddd, ^3J = 8.5 Hz, ^3J = 6.9 Hz, ^4J = 1.5 Hz, 1H, Ar-H), 7.72 (ddd, ^3J = 8.1 Hz, ^3J = 6.8 Hz, ^4J = 1.2 Hz, 1H, Ar-H), 2.75 (s, 1H, C*H*), 2.27 (s, 6H, 3 × C*H$_2$*) ppm; ^{13}C NMR (100 MHz, CDCl$_3$): δ = 156.0 (C$_q$, C_{Ar}SO$_2$), 147.5 (C$_q$, C_{Ar}), 138.7 (+, CH_{Ar}), 131.2 (+, CH_{Ar}), 130.6 (+, CH_{Ar}), 129.4 (+, CH_{Ar}), 129.2 (C$_q$, C_{Ar}), 128.0 (+,

CH_{Ar}), 118.6 (+, CH_{Ar}), 54.5 (C_q, CH_2CS), 51.6 (−, 3 × CCH_2CH), 27.7 (+, CH) ppm; IR (ATR): $\tilde{v}$ = 3109, 2997, 2972, 2919, 2884, 1578, 1497, 1299, 1289, 1204, 1166, 1113, 1088, 878, 832, 766, 674, 649, 612, 588, 547, 520, 476, 445, 399, 388 cm^{-1}; MS (EI, 100 °C): m/z (%) = 259 (0.6) [M]$^+$, 258 (2) [M−H]$^+$,194 (73) [M−C_5H_7+2H]$^+$, 129 (100) [M−$C_5H_7SO_2$+H]$^+$, 128 (73) [M−$C_5H_7SO_2$]$^+$, 67 (18) [C_5H_7]$^+$; HRMS (EI, 100 °C): calc. for $C_{14}H_{13}O_2N^{32}S$ [M]$^+$: 259.0667; found 259.0666.

1-(Bicyclo[1.1.1]pentan-1-ylsulfonyl)isoquinoline (**149f**)

BCP-SO$_2$Na (**169**, 123 mg, 795 µmol, 1.30 equiv.), 1-chloroisoquinoline (**170e**, 100 mg, 611 µmol, 1.00 equiv.) and potassium carbonate (127 mg, 917 µmol, 1.50 equiv.) were mixed in a dry vial under argon atmosphere. DMF (1.0 mL) was added and the mixture was stirred at 120 °C for 16 h. The reaction mixture was cooled down and purified *via* column chromatography (*c*hex/EtOAc 10:1). The product was obtained as a colorless solid (65 mg, 251 µmol, 41%).

R_f (chex/EtOAc 5:1): 0.16; ^{1}H NMR (400 MHz, CDCl$_3$): δ = 8.99 (dq, ^3J = 8.5 Hz, ^4J = 1.0 Hz, 1H, Ar-H), 8.57 (d, ^3J = 5.5 Hz, 1H, Ar-H), 7.93–7.89 (m, 1H, Ar-H), 7.85 (dd, ^3J = 5.5 Hz, ^4J = 0.9 Hz, 1H, Ar-H), 7.77 (ddd, ^3J = 8.2 Hz, ^3J = 6.9 Hz, ^4J = 1.3 Hz, 1H, Ar-H), 7.71 (ddd, ^3J = 8.4 Hz, ^3J = 6.9 Hz, ^4J = 1.5 Hz, 1H, Ar-H), 2.74 (s, 1H, CH), 2.29 (s, 6H, 3 × CH_2) ppm; ^{13}C NMR (100 MHz, CDCl$_3$): δ = 155.3 (C_q, $C_{Ar}SO_2$), 140.8 (+, CH_{Ar}), 137.6 (C_q, 2 × C_{Ar}), 131.3 (+, CH_{Ar}), 129.3 (+, CH_{Ar}), 127.6 (+, CH_{Ar}), 125.4 (+, CH_{Ar}), 125.4 (+, CH_{Ar}), 55.4 (C_q, CH_2CS), 51.8 (−, 3 × CCH_2CH), 27.6 (+, CH) ppm; IR (ATR): $\tilde{v}$ = 3009, 2966, 2918, 1622, 1584, 1557, 1496, 1453, 1373, 1292, 1204, 1160, 1108, 1095, 1020, 993, 949, 874, 840, 827, 786, 773, 748, 683, 656, 613, 569, 518, 462, 435, 408 cm^{-1}; MS (EI, 70 °C): m/z (%) = 259 (3) [M]$^+$, 194 (86) [M−C_5H_7+2H]$^+$, 129 (100) [M−$C_5H_7SO_2$+H]$^+$, 128 (62) [M−$C_5H_7SO_2$]$^+$, 67 (11) [C_5H_7]$^+$; HRMS (EI, 70 °C): calc. for $C_{14}H_{13}O_2N^{32}S$ [M]$^+$: 259.0667; found 259.0666.

4-Nitrophenyl phenyl sulfone (**172**)

Sodium benzenesulfinate (**171**, 151 mg, 921 µmol, 1.30 equiv.) and 1-fluoro-4-nitrobenzene (**170a**, 75 µL, 100 mg, 709 µmol, 1.00 equiv.) were mixed in a dry vial under argon atmosphere. DMF (1.0 mL) was added and the mixture was stirred at 80 °C for 2 h. The reaction mixture was cooled down and purified *via* column chromatography (*c*hex/EtOAc 5:1). The product was obtained as a white solid (180 mg, 684 µmol, 96%).

R_f (chex/EtOAc 5:1): 0.25; ^{1}H NMR (400 MHz, CDCl$_3$): δ = 8.37–8.31 (m, 2H, Ar-H), 8.16–8.10 (m, 2H, Ar-H), 8.00–7.94 (m, 2H, Ar-H), 7.68–7.60 (m, 1H, Ar-H), 7.60–7.52 (m, 2H, Ar-H).

The analytical data is in accordance with the literature.[165] Crystallographic information of the product can be found in chapter 7.4.

Bicyclo[1.1.1]pentyl 4-methoxyphenyl sulfone (**149g**)

 BCP-SO$_2$Na (**169**, 80 mg, 519 µmol, 1.30 equiv.), 1-bromo-4-methoxybenzene (**170f**, 50 µL, 74.5 mg, 398 µmol, 1.00 equiv.), copper(I) iodide (7.6 mg, 40.0 µmol, 0.100 equiv.), L-proline (9.2 mg, 80.0 µmol, 0.200 equiv.) and potassium carbonate (55 mg, 398 µmol, 1.00 equiv.) were mixed in a dry vial under argon atmosphere. DMSO (1.0 mL) was added and the mixture was stirred at 110 °C for 22 h. The reaction mixture was cooled down and purified *via* column chromatography (chex/EtOAc 10:1 to 3:1). The product was obtained as a pale brown oil that slowly solidified (69 mg, 290 µmol, 73%).

R_f (chex/EtOAc 5:1): 0.21; m.p. 64–66 °C; ^{1}H NMR (400 MHz, CDCl$_3$): δ = 7.80–7.75 (m, 2H, Ar-H), 7.04–6.99 (m, 2H, Ar-H), 3.89 (s, 3H, OCH$_3$), 2.71 (s, 1H, CH), 2.06 (s, 6H, 3 × CH$_2$) ppm; ^{13}C NMR (100 MHz, CDCl$_3$): δ = 163.8 (C$_q$, C_{Ar}OCH$_3$), 130.9 (+, 2 × CH$_{Ar}$), 128.5 (C$_q$, C_{Ar}SO$_2$), 114.5 (+, 2 × CH$_{Ar}$), 55.8 (+, OCH$_3$), 55.4 (C$_q$, CH$_2$CS), 50.5 (−, 3 × CCH$_2$CH), 26.8 (+, CH) ppm; IR (ATR): $\tilde{v}$ = 3009, 2966, 2918, 1622, 1584, 1557, 1496, 1453, 1373, 1292, 1204, 1160, 1108, 1095, 1020, 993, 949, 874, 840, 827, 786, 773, 748, 683, 656, 613, 569, 518, 462, 435, 408 cm^{-1}; MS (EI, 60 °C): m/z (%) = 238 (12) [M]$^+$, 172 (36) [M–C$_5$H$_7$+H]$^+$, 155 (100) [M–C$_5$H$_7$–O]$^+$, 67 (55) [C$_5$H$_7$]$^+$; HRMS (EI, 70 °C): calc. for C$_{12}$H$_{14}$O$_3$^{32}S [M]$^+$: 238.0664; found 238.0664.

Bicyclo[1.1.1]pentyl 2-bromophenyl sulfone (**149s**)

BCP-SO$_2$Na (**169**, 80 mg, 519 µmol, 1.30 equiv.), 1-bromo-2-iodobenzene (**170g**, 51 µL, 112 mg, 397 µmol, 1.00 equiv.), copper(I) iodide (7.6 mg, 40.0 µmol, 0.100 equiv.), L-proline (9.2 mg, 80.0 µmol, 0.200 equiv.) and potassium carbonate (55 mg, 398 µmol, 1.00 equiv.) were mixed in a dry vial under argon atmosphere. DMSO (1.0 mL) was added and the mixture was stirred at 100 °C for 18 h. The reaction mixture was cooled down and purified *via* column chromatography (chex/EtOAc 10:1 to 5:1). The product was obtained as a colorless oil that slowly solidified (21 mg, 73.1 µmol, 18%).

Analytical data for **149s** can be found in chapter 7.2.4.

Bicyclo[1.1.1]pentyl methyl sulfone (**149h**)

BCP-SO$_2$Na (**169**, 100 mg, 649 µmol, 1.00 equiv.) was dissolved in DMF (1.0 mL) and stirred at 0 °C (ice bath). A freshly prepared 0.5 M solution of iodomethane in DMF (**173**, 1.95 mL, 138 mg, 973 µmol, 1.50 equiv.) was added dropwise and the reaction mixture was stirred for 4 h at 0 °C. The mixture was diluted with 2 mL of water and extracted with 15 mL of CH$_2$Cl$_2$. The organic layer was collected and was dried by the addition of Na$_2$SO$_4$. The mixture was filtered through a glass funnel and the solvent was evaporated under reduced pressure. The crude product was recrystallized from *c*hex and dried under reduced pressure. The product was obtained as fine, colorless needles (57 mg, 390 µmol, 60%).

m.p. 66–68 °C; ^{1}H NMR (400 MHz, CDCl$_3$): δ = 2.81 (s, 3H, C*H$_3$*), 2.79 (s, 1H, C*H*), 2.26 (s, 6H, 3 × C*H$_2$*) ppm; ^{13}C NMR (100 MHz, CDCl$_3$): δ = 54.5 (C$_q$, *C*SO$_2$CH$_3$), 50.7 (–, 3 × *C*CH$_2$CH), 37.4 (+, *C*H$_3$), 26.4 (+, *C*H) ppm; IR (ATR): $\tilde{v}$ = 3013, 2999, 2979, 2927, 2890, 1741, 1507, 1453, 1411, 1322, 1279, 1208, 1167, 1103, 969, 943, 875, 782, 749, 554, 540, 494 cm^{-1}; MS (EI, 25 °C): *m/z* (%) = 147 (1) [M+H]$^+$, 67 (100) [M–SO$_2$CH$_3$]$^+$; HRMS (EI, 25 °C): calc. for C$_6$H$_{11}$O$_2$^{32}S [M+H]$^+$: 147.0480; found 147.0479.

Crystallographic information of the product can be found in chapter 7.4.

N-Benzyl-*N*-methylbicyclo[1.1.1]pentanesulfonamide (**176a**)

BCP-SO$_2$Na (**169**, 100 mg, 649 µmol, 1.00 equiv.) was dissolved in water (1.0 mL), *N*-benzylmethylamine (**175a**, 126 µL, 118 mg, 973 µmol, 1.50 equiv.) was added, followed by iodine (165 mg, 649 µmol, 1.00 equiv.). The reaction mixture was stirred for 48 h at room temperature, then poured into a separation funnel and diluted with 10 mL of CH$_2$Cl$_2$. The organic layer was washed successively with 10 mL of Na$_2$S$_2$O$_3$-solution. The aqueous layers were recombined and reextracted with CH$_2$Cl$_2$. The organic layers were collected and dried by the addition of Na$_2$SO$_4$. The mixture was filtered through a glass funnel and the solvent was evaporated under reduced pressure. The crude product was purified *via* column chromatography (*c*hex/EtOAc 10:1). The product was obtained as a colorless solid (40 mg, 159 µmol, 25%).

R_f (*c*hex/EtOAc 5:1): 0.26; m.p. 94–96 °C; ^{1}H NMR (400 MHz, CDCl$_3$): δ = 7.37–7.28 (m, 5H, Ar-H), 4.37 (s, 2H, C*H$_2$*NH), 2.81 (s, 3H, C*H$_3$*), 2.73 (s, 1H, C*H*), 2.28 (s, 6H, 3 × CC*H$_2$*CH) ppm; ^{13}C NMR (100 MHz, CDCl$_3$): δ = 136.4 (C$_q$, *C*$_{Ar}$), 128.8 (+, 2 × *C*H$_{Ar}$), 128.4 (+, 2 × *C*H$_{Ar}$), 128.0 (+, *C*H$_{Ar}$), 54.3 (–, NH*C*H$_2$), 53.8 (C$_q$, *C*CH$_2$CH), 52.1 (–, 3 × *C*CH$_2$CH), 34.9 (+, *C*H$_3$), 27.7 (+,

CH) ppm; IR (ATR): $\tilde{v}$ = 2979, 2918, 2885, 1494, 1455, 1373, 1313, 1211, 1197, 1181, 1145, 1113, 1075, 1000, 939, 902, 877, 779, 761, 727, 694, 613, 591, 554, 528, 490, 458 cm^{-1}; MS (ESI+): m/z (%) = 252 (28) [M+H]$^+$, 188 (100) [M–SO$_2$+H]$^+$; HRMS (ESI+): calc. for $C_{13}H_{18}NO_2{}^{32}S$ [M+H]$^+$: 252.1053; found 252.1050.

N-Benzylbicyclo[1.1.1]pentanesulfonamide (**176b**)

BCP-SO$_2$Na (**169**, 100 mg, 649 μmol, 1.00 equiv.) was dissolved in water (1.0 mL), benzylamine (**175b**, 92 μL, 90.4 mg, 843 μmol, 1.30 equiv.) was added, followed by iodine (165 mg, 649 μmol, 1.00 equiv.). The reaction mixture was stirred for 48 h at room temperature, then poured into a separation funnel and diluted with 10 mL of CH$_2$Cl$_2$. The organic layer was washed successively with 10 mL of Na$_2$S$_2$O$_3$-solution. The aqueous layers were recombined and reextracted with CH$_2$Cl$_2$. The organic layers were collected and dried by the addition of Na$_2$SO$_4$. The mixture was filtered through a glass funnel and the solvent was evaporated under reduced pressure. The crude product was purified *via* column chromatography (*c*hex/EtOAc 10:1 to 5:1). The product was obtained as a colorless solid (38 mg, 160 μmol, 25%).

R_f (*c*hex/EtOAc 5:1): 0.18; m.p. 116–118 °C; ^{1}H NMR (400 MHz, CDCl$_3$): δ = 7.39–7.28 (m, 5H, Ar-H), 4.49 (bs, 1H, NH), 4.31 (d, ^3J = 6.1 Hz, 2H, CH_2NH), 2.68 (s, 1H, CH), 2.17 (s, 6H, 3 × CCH_2CH) ppm; ^{13}C NMR (100 MHz, CDCl$_3$): δ = 137.5 (C$_q$, C_{Ar}), 129.0 (+, 2 × CH$_{Ar}$), 128.1 (+, 2 × CH$_{Ar}$), 127.9 (+, CH$_{Ar}$), 54.3 (C$_q$, CCH$_2$CH), 51.3 (–, 3 × CCH$_2$CH), 47.6 (–, NHCH$_2$), 26.7 (+, CH) ppm; IR (ATR): $\tilde{v}$ = 3286, 3247, 2978, 2918, 2880, 1494, 1452, 1439, 1431, 1302, 1211, 1183, 1146, 1120, 1082, 1062, 1051, 1026, 908, 878, 850, 829, 748, 724, 693, 662, 620, 588,   545,   527,   509, 452 cm^{-1}; MS (ESI+): m/z (%) = 260 (9) [M+Na]$^+$, 238 (6) [M+H]$^+$, 174 (100) [M–SO$_2$+H]$^+$; HRMS (ESI+): calc. for $C_{12}H_{16}NO_2{}^{32}S$ [M+H]$^+$: 238.0896; found 238.0894.

Crystallographic information of the product can be found in chapter 7.4.

Bicyclo[1.1.1]pentanesulfonamide (**176c**)

BCP-SO$_2$Na (**169**, 1.00 g, 6.50 mmol, 1.00 equiv.) was dissolved in water (30 mL) and hydroxylamine-O-sulfonic acid (1.47 g, 13.0 mmol, 2.00 equiv.) and potassium acetate (637 mg, 6.50 mmol, 1.00 equiv.) were added. The reaction mixture was stirred for 18 h at room temperature. Then, the solution was basified with 4 M NaOH-solution and extracted with EtOAc (3 × 50 mL). The organic layers were collected and were dried by the addition of Na$_2$SO$_4$.

The mixture was filtered through a glass funnel and the solvent was evaporated under reduced pressure. The product was obtained as a colorless solid (821 mg, 5.58 mmol, 86%).

m.p. could not be determined, product decomposed; ^{1}H NMR (500 MHz, DMSO-d_6): δ = 6.76 (bs, 2H, NH_2), 2.67 (s, 1H, CH), 2.04 (s, 6H, 3 × CH_2) ppm; ^{13}C NMR (125 MHz, DMSO-d_6): δ = 54.4 (C$_q$, CSO$_2$NH$_2$), 50.0 (–, 3 × CCH$_2$CH), 25.2 (+, CH) ppm; IR (ATR): $\tilde{v}$ = 3302, 3218, 3002, 2975, 2922, 2885, 2853, 1738, 1459, 1295, 1208, 1200, 1170, 1111, 1074, 945, 921, 877, 721, 673, 628, 581, 548, 499, 453 cm^{-1}; MS (ESI+): m/z (%) = 170 (100) [M+Na]$^+$; HRMS (ESI+): calc. for C$_5$H$_9$NO$_2$^{32}S [M+Na]$^+$: 170.0246; found 170.0246.

Bicyclo[1.1.1]pentylsulfinic chloride (**178**)

BCP-SO$_2$Na (**169**, 154 mg, 1.00 mmol, 1.00 equiv.) was suspended in CH$_2$Cl$_2$ (1.5 mL) and stirred at room temperature. Thionyl chloride (109 µL, 179 mg, 1.50 mmol, 1.50 equiv.) was added, followed by catalytic DMF (8 µL, 7.55 mg, 100 µmol, 0.100 equiv.) and the reaction mixture was stirred for 16 h at room temperature. The crude solution was used directly in the next step.

Bicyclo[1.1.1]pentyl phenyl sulfoxide (**107a**)

The crude solution of the sulfinyl chloride (**178**, 500 µL, 333 µmol) was cooled to –78 °C (dry ice/acetone bath) and a 1 M phenylmagnesium bromide solution (834 µL, 151 mg, 834 µmol, 2.50 equiv.) in THF was added. The reaction mixture was allowed to warm to room temperature over 1 h. The reaction was quenched by careful addition of 1 mL of sat. NH$_4$Cl-solution, extracted with CH$_2$Cl$_2$ (2 × 10 mL), dried by the addition of Na$_2$SO$_4$, filtered and evaporated. The crude product was purified *via* column chromatography (*c*hex/EtOAc 98:2 to 65:35). The product was obtained as a colorless oil (38 mg, 198 µmol, 59%).

Analytical data for **107a** can be found in chapter 7.2.4.

Bicyclo[1.1.1]pentyl 4-chlorophenyl sulfoxide (**107b**)

The crude solution of the sulfinyl chloride (**178**, 850 µL, 567 µmol) was cooled to –78 °C (dry ice/acetone bath) and a 1 M 4-chlorophenylmagnesium bromide solution (1.14 mL, 246 mg, 1.14 mmol, 2.01 equiv.) in Et$_2$O was added. The reaction mixture was allowed to warm to room temperature over 1 h. The reaction was quenched by careful addition of 1 mL of sat. NH$_4$Cl-solution, extracted with CH$_2$Cl$_2$ (2 × 10 mL), dried by the addition of Na$_2$SO$_4$, filtered and evaporated. The crude product was purified *via* column chromatography (*c*hex/EtOAc 4:1). The product was obtained as a pale yellow oil (45 mg, 199 μmol, 35%).

R_f (*c*hex/EtOAc 4:1): 0.12; ^{1}H NMR (400 MHz, CDCl$_3$): δ = 7.47–7.51 (m, 2H, Ar-H), 7.43–7.47 (m, 2H, Ar-H), 2.84 (s, 1H, C*H*), 1.85–1.92 (m, 6H, 3 × C*H$_2$*) ppm; ^{13}C NMR (100 MHz, CDCl$_3$): δ = 140.3 (C$_q$, *C*$_{Ar}$S), 137.1 (C$_q$, *C*$_{Ar}$Cl), 129.4 (+, 2 × *C*H$_{Ar}$), 125.7 (+, 2 × *C*H$_{Ar}$), 55.4 (C$_q$, CH$_2$*C*S), 49.0 (–, 3 × C*C*H$_2$CH), 27.9 (+, *C*H) ppm; IR (ATR): v = 2990, 1916, 2879, 1574, 1474, 1389, 1202, 1129, 1079, 1040, 1010, 882, 868, 823, 773, 740, 558, 510, 444 cm^{-1}; MS (EI, 60 °C): *m/z* (%) = 226 (1) [M]$^+$, 160 (89) [M–C$_5$H$_7$+H]$^+$, 159 (17) [M–C$_5$H$_7$]$^+$, 142 (2) [M-C$_5$H$_7$O]$^+$, 112 (27), 111 (14) [M-C$_5$H$_7$OS]$^+$, 99 (3) [M-C$_6$H$_4$ClO]$^+$, 76 (4) [M-C$_6$H$_4$]$^+$, 67 (100) [C$_5$H$_7$]$^+$, 65 (30); HRMS (EI, 60 °C): calc. for C$_{11}$H$_{11}$ClO^{32}S [M]$^+$: 226.0219, found 226.0217.

7.3   Selected NMR spectra

Control experiment in Scheme 25: ¹H NMR

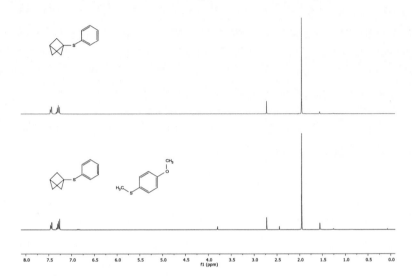

Side-product **117** in Scheme 26: ¹H NMR

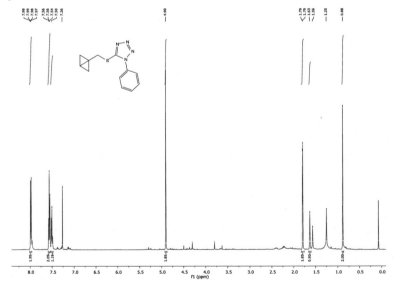

¹³C NMR

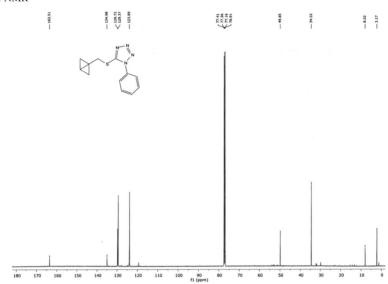

HSQC

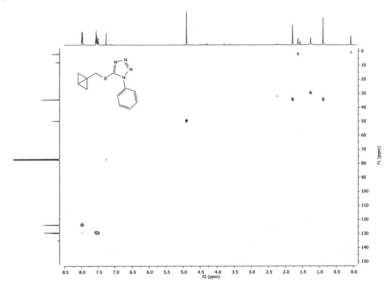

HMBC

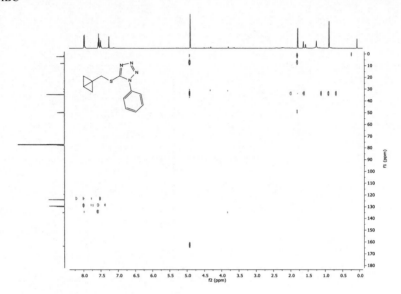

Competitive reaction in Scheme 30: ¹H NMR

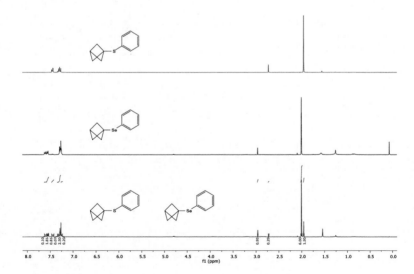

Mixture of isomers **105e**: ^{1}H NMR

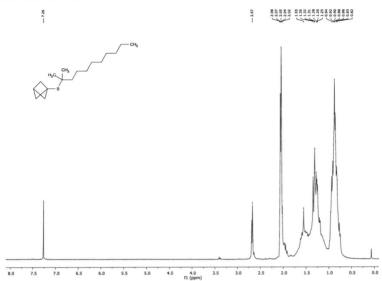

Mixture study in Scheme 33: ^{1}H NMR

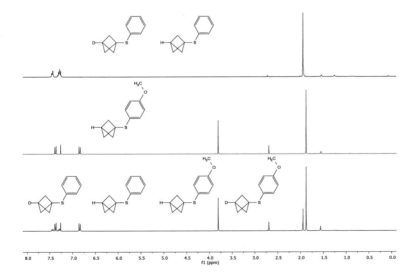

Radical scavenger reaction in Scheme 34: ^{1}H NMR

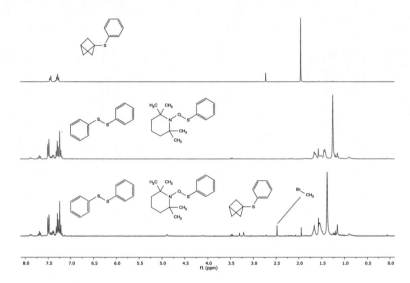

Product **106d** mixed with disulfide **138d**: ^{1}H NMR

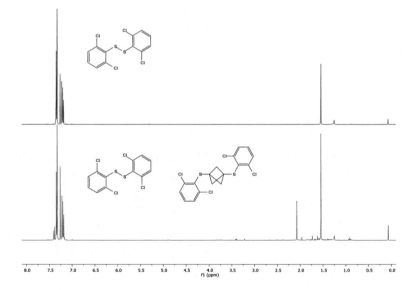

Product mixture in Scheme 41: ^{1}H NMR

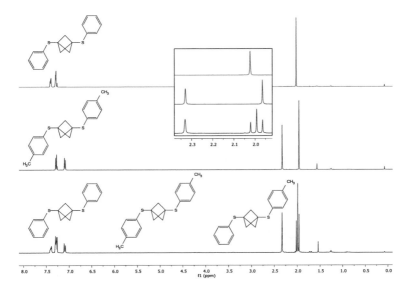

Comparison of products from Scheme 46: ^{1}H NMR

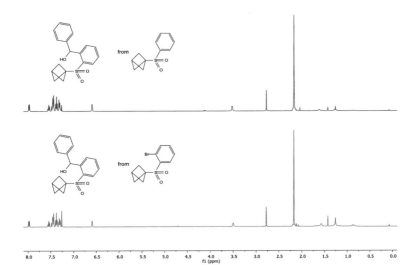

^{13}C NMR

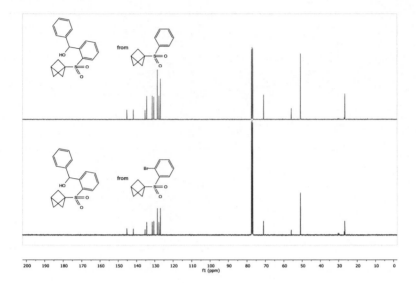

7.4 Crystallographic data

7.4.1 Crystallographic data solved by Dr. Martin Nieger

Crystal structures in this section were measured and solved by Dr. Martin Nieger at the University of Helsinki (Finland).

Table 11. Measured and solved crystal structures by Dr. Martin Nieger.

Numbering in this thesis	Sample code used by Dr. Martin Nieger	CCDC number (if available)
104k	sb1030	1575330
104o	sb1041	1575329
138g	sb1197	–
106a	sb1083	1896780
106k	sb1190	1908215
108a	sb1084	1951387
108f	sb1170	1951390
149b	sb1179	1959595
172	sb1217	–
149c	sb1210	1959596
149d	sb1204	1959597
149h	sb1214	1959598
176b	sb1209	1959599

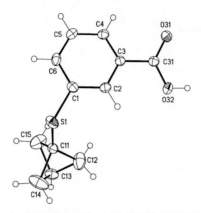

3-(bicyclo[1.1.1]pentan-1-yl-thio)benzoic acid (104k) SB1030_HY

Crystal data

$C_{12}H_{12}O_2S$	$F(000) = 464$
$M_r = 220.28$	$D_x = 1.333$ Mg m^{-3}
Monoclinic, $P2_1/n$ *(no.14)*	Cu $K\alpha$ radiation, $\lambda = 1.54178$ Å
$a = 12.6580$ (4) Å	Cell parameters from 4608 reflections
$b = 5.5459$ (2) Å	$\theta = 4.4$–$72.1°$
$c = 15.6396$ (5) Å	$\mu = 2.43$ mm^{-1}
$\beta = 90.510$ (2)°	$T = 123$ K
$V = 1097.86$ (6) Å^3	Plates, colorless
$Z = 4$	$0.20 \times 0.06 \times 0.02$ mm

Data collection

Bruker D8 VENTURE diffractometer with Photon100 detector	2161 independent reflections
Radiation source: INCOATEC microfocus sealed tube	1799 reflections with $I > 2\sigma(I)$
Detector resolution: 10.4167 pixels mm^{-1}	$R_{int} = 0.056$
rotation in ϕ and ω, 1°, shutterless scans	$\theta_{max} = 72.4°$, $\theta_{min} = 4.5°$
Absorption correction: multi-scan *SADABS* (Sheldrick, 2014)	$h = -15{\rightarrow}15$
$T_{min} = 0.824$, $T_{max} = 0.958$	$k = -6{\rightarrow}6$
10208 measured reflections	$l = -19{\rightarrow}19$

Refinement

Refinement on F^2	Primary atom site location: dual
Least-squares matrix: full	Secondary atom site location: difference Fourier map
$R[F^2 > 2\sigma(F^2)] = 0.046$	Hydrogen site location: difference Fourier map
$wR(F^2) = 0.111$	H atoms treated by a mixture of independent and constrained refinement
$S = 1.08$	$w = 1/[\sigma^2(F_o^2) + (0.0467P)^2 + 0.9117P]$ where $P = (F_o^2 + 2F_c^2)/3$
2161 reflections	$(\Delta/\sigma)_{max} < 0.001$
139 parameters	$\Delta\rangle_{max} = 0.38$ e Å^{-3}
1 restraint	$\Delta\rangle_{min} = -0.25$ e Å^{-3}

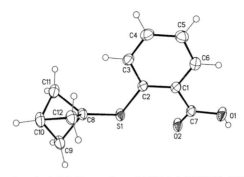

2-(bicyclo[1.1.1]pentan-1-yl-thio)benzoic acid (104o) SB1041_HY

Crystal data

$C_{12}H_{12}O_2S$	$Z = 2$
$M_r = 220.28$	$F(000) = 232$
Triclinic, $P\text{-}1$ (no.2)	$D_x = 1.385$ Mg m^{-3}
$a = 7.9867$ (4) Å	Cu $K\alpha$ radiation, $\lambda = 1.54178$ Å
$b = 8.1535$ (4) Å	Cell parameters from 4019 reflections
$c = 8.4570$ (4) Å	$\theta = 5.2$–72.0°
$\alpha = 86.927$ (3)°	$\mu = 2.52$ mm^{-1}
$\beta = 83.840$ (3)°	$T = 123$ K
$\gamma = 74.768$ (2)°	Plates, colorless
$V = 528.12$ (5) Å^3	0.08 × 0.08 × 0.02 mm

Data collection

Bruker D8 VENTURE diffractometer with Photon100 detector	2046 independent reflections
Radiation source: INCOATEC microfocus sealed tube	1867 reflections with $I > 2\sigma(I)$
Detector resolution: 10.4167 pixels mm^{-1}	$R_{int} = 0.027$
rotation in ϕ and ω, 1°, shutterless scans	$\theta_{max} = 72.0°$, $\theta_{min} = 5.3°$
Absorption correction: multi-scan *SADABS* (Sheldrick, 2014)	$h = -9 \rightarrow 9$
$T_{min} = 0.814$, $T_{max} = 0.958$	$k = -10 \rightarrow 10$
5082 measured reflections	$l = -10 \rightarrow 10$

Refinement

Refinement on F^2	Primary atom site location: dual
Least-squares matrix: full	Secondary atom site location: difference Fourier map
$R[F^2 > 2\sigma(F^2)] = 0.035$	Hydrogen site location: difference Fourier map
$wR(F^2) = 0.091$	H atoms treated by a mixture of independent and constrained refinement
$S = 1.04$	$w = 1/[\sigma^2(F_o^2) + (0.0451P)^2 + 0.2715P]$ where $P = (F_o^2 + 2F_c^2)/3$
2046 reflections	$(\Delta/\sigma)_{max} = 0.001$
139 parameters	$\Delta\rangle_{max} = 0.53$ e Å^{-3}
1 restraint	$\Delta\rangle_{min} = -0.29$ e Å^{-3}

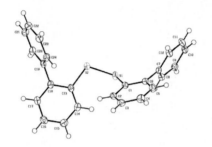

1,2-di([1,1'-biphenyl]-2-yl)disulfane (138g) SB1197_HY

Crystal data

$C_{24}H_{18}S_2$	$F(000) = 776$
$M_r = 370.50$	$D_x = 1.306$ Mg m^{-3}
Monoclinic, $P2_1/c$ (no.14)	Cu $K\alpha$ radiation, $\lambda = 1.54178$ Å
$a = 13.7844$ (5) Å	Cell parameters from 9813 reflections
$b = 10.3009$ (3) Å	$\theta = 5.3$–$72.1°$
$c = 13.5123$ (4) Å	$\mu = 2.57$ mm^{-1}
$\beta = 100.748$ (1)°	$T = 123$ K
$V = 1884.98$ (10) Å^3	Blocks, colorless
$Z = 4$	$0.24 \times 0.18 \times 0.12$ mm

Data collection

Bruker D8 VENTURE diffractometer with PhotonII CPAD detector	3662 reflections with $I > 2\sigma(I)$
Radiation source: INCOATEC microfocus sealed tube	$R_{int} = 0.025$
rotation in ϕ and ω, 1°, shutterless scans	$\theta_{max} = 72.2°$, $\theta_{min} = 3.3°$
Absorption correction: multi-scan *SADABS* (Sheldrick, 2014)	$h = -16 \rightarrow 17$
$T_{min} = 0.669$, $T_{max} = 0.774$	$k = -12 \rightarrow 12$
24725 measured reflections	$l = -16 \rightarrow 16$
3696 independent reflections	

Refinement

Refinement on F^2	Secondary atom site location: difference Fourier map
Least-squares matrix: full	Hydrogen site location: difference Fourier map
$R[F^2 > 2\sigma(F^2)] = 0.028$	H-atom parameters constrained
$wR(F^2) = 0.074$	$w = 1/[\sigma^2(F_o^2) + (0.038P)^2 + 0.764P]$ where $P = (F_o^2 + 2F_c^2)/3$
$S = 1.04$	$(\Delta/\sigma)_{max} = 0.001$
3696 reflections	$\Delta\rangle_{max} = 0.27$ e Å^{-3}
236 parameters	$\Delta\rangle_{min} = -0.29$ e Å^{-3}
0 restraints	Extinction correction: *SHELXL2014/7* (Sheldrick 2014), Fc*=kFc[1+0.001xFc$^2\lambda^3$/sin(2θ)]$^{-1/4}$
Primary atom site location: dual	Extinction coefficient: 0.00145 (27)

1,3-bis(phenylthio)bicyclo[1.1.1]pentane (106a) SB1083_HY

Crystal data

$C_{17}H_{16}S_2$	$F(000) = 300$
$M_r = 284.42$	$D_x = 1.282$ Mg m^{-3}
Monoclinic, $P2_1$ (no.4)	Cu $K\alpha$ radiation, $\lambda = 1.54178$ Å
$a = 9.7073$ (3) Å	Cell parameters from 9968 reflections
$b = 5.6219$ (2) Å	$\theta = 3.2$–$72.0°$
$c = 14.3212$ (5) Å	$\mu = 3.12$ mm^{-1}
$\beta = 109.485$ (1)$°$	$T = 123$ K
$V = 736.80$ (4) Å^3	Plates, colorless
$Z = 2$	$0.36 \times 0.16 \times 0.06$ mm

Data collection

Bruker D8 VENTURE diffractometer with PhotonII CPAD detector	2763 reflections with $I > 2\sigma(I)$
Radiation source: INCOATEC microfocus sealed tube	$R_{int} = 0.030$
rotation in ϕ and ω, 1°, shutterless scans	$\theta_{max} = 72.1°$, $\theta_{min} = 3.3°$
Absorption correction: multi-scan SADABS (Sheldrick, 2014)	$h = -11 \rightarrow 11$
$T_{min} = 0.509$, $T_{max} = 0.754$	$k = -6 \rightarrow 5$
14308 measured reflections	$l = -17 \rightarrow 17$
2772 independent reflections	

Refinement

Refinement on F^2	Secondary atom site location: difference Fourier map
Least-squares matrix: full	Hydrogen site location: difference Fourier map
$R[F^2 > 2\sigma(F^2)] = 0.026$	H-atom parameters constrained
$wR(F^2) = 0.070$	$w = 1/[\sigma^2(F_o^2) + (0.0517P)^2 + 0.0998P]$ where $P = (F_o^2 + 2F_c^2)/3$
$S = 1.04$	$(\Delta/\sigma)_{max} = 0.001$
2772 reflections	$\Delta\rangle_{max} = 0.32$ e Å^{-3}
172 parameters	$\Delta\rangle_{min} = -0.17$ e Å^{-3}
1 restraint	Absolute structure: Flack x determined using 1171 quotients [(I+)-(I-)]/[(I+)+(I-)] (Parsons, Flack and Wagner, Acta Cryst. B69 (2013) 249-259).
Primary atom site location: dual	Absolute structure parameter: -0.011 (12)

(4-methoxyphenyl)(3-(phenylthio)bicyclo-[1.1.1]pentan-1-yl)sulfane   (106k)
SB1190_HY

Crystal data

$C_{18}H_{18}OS_2$	$F(000) = 332$
$M_r = 314.44$	$D_x = 1.324$ Mg m^{-3}
Monoclinic, $P2_1$ (no.4)	Cu $K\alpha$ radiation, $\lambda = 1.54178$ Å
$a = 9.6338$ (5) Å	Cell parameters from 9874 reflections
$b = 5.6921$ (3) Å	$\theta = 3.0–72.2°$
$c = 14.5108$ (7) Å	$\mu = 3.01$ mm^{-1}
$\beta = 97.628$ (1)°	$T = 123$ K
$V = 788.68$ (7) Å^3	Blocks, colorless
$Z = 2$	$0.32 \times 0.16 \times 0.08$ mm

Data collection

Bruker D8 VENTURE diffractometer with PhotonII CPAD detector	3059 reflections with $I > 2\sigma(I)$
Radiation source: INCOATEC microfocus sealed tube	$R_{int} = 0.027$
rotation in ϕ and ω, 1°, shutterless scans	$\theta_{max} = 72.2°$, $\theta_{min} = 3.1°$
Absorption correction: multi-scan SADABS (Sheldrick, 2014)	$h = -11 \rightarrow 11$
$T_{min} = 0.665$, $T_{max} = 0.795$	$k = -7 \rightarrow 7$
13068 measured reflections	$l = -16 \rightarrow 17$
3071 independent reflections	

Refinement

Refinement on F^2	Secondary atom site location: difference Fourier map
Least-squares matrix: full	Hydrogen site location: difference Fourier map
$R[F^2 > 2\sigma(F^2)] = 0.024$	H-atom parameters constrained
$wR(F^2) = 0.063$	$w = 1/[\sigma^2(F_o^2) + (0.0418P)^2 + 0.1312P]$ where $P = (F_o^2 + 2F_c^2)/3$
$S = 1.04$	$(\Delta/\sigma)_{max} < 0.001$
3071 reflections	$\Delta_{max} = 0.31$ e Å^{-3}
191 parameters	$\Delta_{min} = -0.14$ e Å^{-3}
1 restraint	Absolute structure: Flack x determined using 1351 quotients [(I+)-(I-)]/[(I+)+(I-)] (Parsons, Flack and Wagner, Acta Cryst. B69 (2013) 249-259).
Primary atom site location: dual	Absolute structure parameter: 0.003 (7)

bicyclo[1.1.1]pentan-1-yl(imino)(phenyl)-λ^6-sulfanone (108a) SB1084_HY

Crystal data

$C_{11}H_{13}NOS$	$D_x = 1.334$ Mg m^{-3}
$M_r = 207.28$	Cu $K\alpha$ radiation, $\lambda = 1.54178$ Å
Orthorhombic, $P2_12_12_1$ (no.19)	Cell parameters from 9578 reflections
$a = 6.4986$ (3) Å	$\theta = 5.2$–72.1°
$b = 9.3449$ (4) Å	$\mu = 2.50$ mm^{-1}
$c = 16.9901$ (8) Å	$T = 123$ K
$V = 1031.79$ (8) Å^3	Blocks, colorless
$Z = 4$	$0.20 \times 0.08 \times 0.04$ mm
$F(000) = 440$	

Data collection

Bruker D8 VENTURE diffractometer with PhotonII CPAD detector	1980 reflections with $I > 2\sigma(I)$
Radiation source: INCOATEC microfocus sealed tube	$R_{int} = 0.040$
rotation in ϕ and ω, 1°, shutterless scans	$\theta_{max} = 72.2°$, $\theta_{min} = 5.2°$
Absorption correction: multi-scan SADABS (Sheldrick, 2014)	$h = -8 \rightarrow 8$
$T_{min} = 0.702$, $T_{max} = 0.889$	$k = -10 \rightarrow 11$
10161 measured reflections	$l = -20 \rightarrow 20$
2013 independent reflections	

Refinement

Refinement on F^2	Secondary atom site location: difference Fourier map
Least-squares matrix: full	Hydrogen site location: difference Fourier map
$R[F^2 > 2\sigma(F^2)] = 0.030$	H atoms treated by a mixture of independent and constrained refinement
$wR(F^2) = 0.077$	$w = 1/[\sigma^2(F_o^2) + (0.0393P)^2 + 0.4117P]$ where $P = (F_o^2 + 2F_c^2)/3$
$S = 1.05$	$(\Delta/\sigma)_{max} < 0.001$
2013 reflections	$\Delta\rangle_{max} = 0.24$ e Å^{-3}
134 parameters	$\Delta\rangle_{min} = -0.32$ e Å^{-3}
2 restraints	Absolute structure: Refined as an inversion twin (BASF = 0.38(2)). Parsons' x determined using 805 quotients [(I+)-(I-)]/[(I+)+(I-)] (Parsons, Flack and Wagner, Acta Cryst. B69 (2013) 249-259), Parsons' x = 0.393(8)
Primary atom site location: dual	Absolute structure parameter: 0.38 (2)

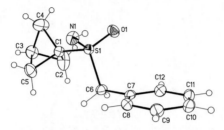

benzyl(bi-cyclo[1.1.1]pentan-1-yl)(imino)-λ^6-sulfanone (108f) SB1170_HY

Crystal data

$C_{12}H_{15}NOS$	$D_x = 1.297$ Mg m^{-3}
$M_r = 221.31$	Cu $K\alpha$ radiation, $\lambda = 1.54178$ Å
Orthorhombic, *Pbcn* (*no.*60)	Cell parameters from 9959 reflections
$a = 22.1804$ (5) Å	$\theta = 3.9–72.0°$
$b = 11.3989$ (3) Å	$\mu = 2.31$ mm^{-1}
$c = 8.9685$ (2) Å	$T = 123$ K
$V = 2267.53$ (9) Å^3	Plates, colorless
$Z = 8$	$0.16 \times 0.12 \times 0.03$ mm
$F(000) = 944$	

Data collection

Bruker D8 VENTURE diffractometer with PhotonII CPAD detector	2072 reflections with $I > 2\sigma(I)$
Radiation source: INCOATEC microfocus sealed tube	$R_{int} = 0.031$
rotation in ϕ and ω, 1°, shutterless scans	$\theta_{max} = 72.1°$, $\theta_{min} = 4.0°$
Absorption correction: multi-scan *SADABS* (Shelsdrick, 2014)	$h = -26 \rightarrow 27$
$T_{min} = 0.767$, $T_{max} = 0.915$	$k = -13 \rightarrow 14$
17304 measured reflections	$l = -11 \rightarrow 9$
2231 independent reflections	

Refinement

Refinement on F^2	Primary atom site location: dual
Least-squares matrix: full	Secondary atom site location: difference Fourier map
$R[F^2 > 2\sigma(F^2)] = 0.037$	Hydrogen site location: difference Fourier map
$wR(F^2) = 0.097$	H atoms treated by a mixture of independent and constrained refinement
$S = 1.07$	$w = 1/[\sigma^2(F_o^2) + (0.0491P)^2 + 1.5136P]$ where $P = (F_o^2 + 2F_c^2)/3$
2231 reflections	$(\Delta/\sigma)_{max} < 0.001$
142 parameters	$\Delta\rangle_{max} = 0.45$ e Å^{-3}
32 restraints	$\Delta\rangle_{min} = -0.33$ e Å^{-3}

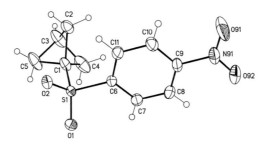

1-((4-nitrophenyl)sulfonyl)bicyclo[1.1.1]pentane (149b) SB1179_HY

Crystal data

$C_{11}H_{11}NO_4S$	$Z = 2$
$M_r = 253.27$	$F(000) = 264$
Triclinic, $P\text{-}1$ (no.2)	$D_x = 1.478$ Mg m^{-3}
$a = 7.3339$ (3) Å	Cu $K\alpha$ radiation, $\lambda = 1.54178$ Å
$b = 7.9774$ (3) Å	Cell parameters from 9939 reflections
$c = 10.7533$ (4) Å	$\theta = 4.5$–72.1°
$\alpha = 68.613$ (1)°	$\mu = 2.59$ mm^{-1}
$\beta = 76.287$ (1)°	$T = 123$ K
$\gamma = 85.342$ (1)°	Blocks, colorless
$V = 569.10$ (4) Å^3	$0.16 \times 0.08 \times 0.06$ mm

Data collection

Bruker D8 VENTURE diffractometer with PhotonII CPAD detector	2186 reflections with $I > 2\sigma(I)$
Radiation source: INCOATEC microfocus sealed tube	$R_{int} = 0.027$
rotation in ϕ and ω, 1°, shutterless scans	$\theta_{max} = 72.1°$, $\theta_{min} = 4.5°$
Absorption correction: multi-scan *SADABS* (Sheldrick, 2014)	$h = -9 \rightarrow 8$
$T_{min} = 0.738$, $T_{max} = 0.841$	$k = -9 \rightarrow 9$
11823 measured reflections	$l = -13 \rightarrow 13$
2222 independent reflections	

Refinement

Refinement on F^2	Primary atom site location: dual
Least-squares matrix: full	Secondary atom site location: difference Fourier map
$R[F^2 > 2\sigma(F^2)] = 0.033$	Hydrogen site location: difference Fourier map
$wR(F^2) = 0.087$	H atoms treated by a mixture of independent and constrained refinement
$S = 1.06$	$w = 1/[\sigma^2(F_o^2) + (0.0437P)^2 + 0.2985P]$ where $P = (F_o^2 + 2F_c^2)/3$
2222 reflections	$(\Delta/\sigma)_{max} = 0.001$
157 parameters	$\Delta\rangle_{max} = 0.37$ e Å^{-3}
1 restraint	$\Delta\rangle_{min} = -0.36$ e Å^{-3}

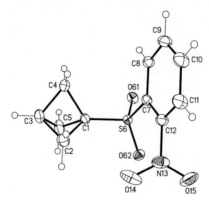

1-((2-nitrophenyl)sulfonyl)bicyclo[1.1.1]pentane (149c) SB1210_HY

Crystal data

$C_{11}H_{11}NO_4S$	$F(000) = 528$
$M_r = 253.27$	$D_x = 1.456$ Mg m^{-3}
Monoclinic, $P2_1/n$ (no.14)	Cu $K\alpha$ radiation, $\lambda = 1.54178$ Å
$a = 9.7936$ (3) Å	Cell parameters from 9967 reflections
$b = 8.6615$ (3) Å	$\theta = 3.2–72.1°$
$c = 13.9543$ (4) Å	$\mu = 2.55$ mm^{-1}
$\beta = 102.614$ (1)°	$T = 123$ K
$V = 1155.13$ (6) Å^3	Blocks, colorless
$Z = 4$	$0.30 \times 0.25 \times 0.20$ mm

Data collection

Bruker D8 VENTURE diffractometer with PhotonII CPAD detector	2237 reflections with $I > 2\sigma(I)$
Radiation source: INCOATEC microfocus sealed tube	$R_{int} = 0.023$
rotation in ϕ and ω, 1°, shutterless scans	$\theta_{max} = 72.2°$, $\theta_{min} = 5.0°$
Absorption correction: multi-scan *SADABS* V2014/5 (Bruker AXS Inc.)	$h = -12 \rightarrow 11$
$T_{min} = 0.525$, $T_{max} = 0.623$	$k = -10 \rightarrow 7$
11641 measured reflections	$l = -17 \rightarrow 17$
2246 independent reflections	

Refinement

Refinement on F^2	Primary atom site location: dual
Least-squares matrix: full	Secondary atom site location: difference Fourier map
$R[F^2 > 2\sigma(F^2)] = 0.029$	Hydrogen site location: difference Fourier map
$wR(F^2) = 0.070$	H atoms treated by a mixture of independent and constrained refinement
$S = 1.05$	$w = 1/[\sigma^2(F_o^2) + (0.0235P)^2 + 0.7517P]$ where $P = (F_o^2 + 2F_c^2)/3$
2246 reflections	$(\Delta/\sigma)_{max} < 0.001$
157 parameters	$\Delta\rangle_{max} = 0.31$ e Å^{-3}
1 restraint	$\Delta\rangle_{min} = -0.36$ e Å^{-3}

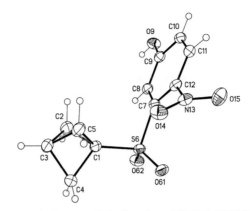

3-(bicyclo[1.1.1]pentan-1-ylsulfonyl)-4-nitrophenol (149d) SB1204_HY

Crystal data

$C_{11}H_{11}NO_5S$	$F(000) = 560$
$M_r = 269.27$	$D_x = 1.557$ Mg m^{-3}
Monoclinic, $P2_1/c$ *(no.14)*	Cu $K\alpha$ radiation, $\lambda = 1.54178$ Å
$a = 8.3279$ (4) Å	Cell parameters from 5602 reflections
$b = 9.6745$ (5) Å	$\theta = 5.5–72.1°$
$c = 14.8135$ (8) Å	$\mu = 2.67$ mm^{-1}
$\beta = 105.753$ (2)°	$T = 123$ K
$V = 1148.67$ (10) Å^3	Plates, colorless
$Z = 4$	$0.08 \times 0.04 \times 0.01$ mm

Data collection

Bruker D8 VENTURE diffractometer with PhotonII CPAD detector	2029 reflections with $I > 2\sigma(I)$
Radiation source: INCOATEC microfocus sealed tube	$R_{int} = 0.040$
rotation in ϕ and ω, 1°, shutterless scans	$\theta_{max} = 72.2°$, $\theta_{min} = 5.5°$
Absorption correction: multi-scan *SADABS* V2014/5 (Bruker AXS Inc.)	$h = -10\rightarrow10$
$T_{min} = 0.823$, $T_{max} = 0.958$	$k = -11\rightarrow11$
9976 measured reflections	$l = -18\rightarrow17$
2251 independent reflections	

Refinement

Refinement on F^2	Primary atom site location: dual
Least-squares matrix: full	Secondary atom site location: difference Fourier map
$R[F^2 > 2\sigma(F^2)] = 0.036$	Hydrogen site location: difference Fourier map
$wR(F^2) = 0.093$	H atoms treated by a mixture of independent and constrained refinement
$S = 1.04$	$w = 1/[\sigma^2(F_o^2) + (0.0505P)^2 + 0.5497P]$ where $P = (F_o^2 + 2F_c^2)/3$
2251 reflections	$(\Delta/\sigma)_{max} = 0.001$
169 parameters	$\Delta\rangle_{max} = 0.60$ e Å^{-3}
2 restraints	$\Delta\rangle_{min} = -0.31$ e Å^{-3}

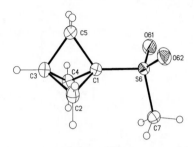

1-(methylsulfonyl)bicyclo[1.1.1]pentane (149h) SB1214_HY

Crystal data

$C_6H_{10}O_2S$	$F(000) = 312$
$M_r = 146.20$	$D_x = 1.375$ Mg m^{-3}
Monoclinic, $P2_1/c$ *(no.14)*	Cu $K\alpha$ radiation, $\lambda = 1.54178$ Å
$a = 5.2606$ (5) Å	Cell parameters from 5793 reflections
$b = 13.1057$ (14) Å	$\theta = 4.3$–$72.0°$
$c = 10.3357$ (11) Å	$\mu = 3.47$ mm^{-1}
$\beta = 97.507$ (4)°	$T = 123$ K
$V = 706.48$ (13) Å^3	Plates, colorless
$Z = 4$	$0.18 \times 0.06 \times 0.02$ mm

Data collection

Bruker D8 VENTURE diffractometer with PhotonII CPAD detector	1361 reflections with $I > 2\sigma(I)$
Radiation source: INCOATEC microfocus sealed tube	$R_{int} = 0.036$
rotation in ϕ and ω, 1°, shutterless scans	$\theta_{max} = 72.1°$, $\theta_{min} = 5.5°$
Absorption correction: multi-scan SADABS V2014/5 (Bruker AXS Inc.)	$h = -6 \rightarrow 6$
$T_{min} = 0.652$, $T_{max} = 0.915$	$k = -16 \rightarrow 13$
6055 measured reflections	$l = -11 \rightarrow 12$
1392 independent reflections	

Refinement

Refinement on F^2	Primary atom site location: dual
Least-squares matrix: full	Secondary atom site location: difference Fourier map
$R[F^2 > 2\sigma(F^2)] = 0.032$	Hydrogen site location: difference Fourier map
$wR(F^2) = 0.084$	H atoms treated by a mixture of independent and constrained refinement
$S = 1.06$	$w = 1/[\sigma^2(F_o^2) + (0.0377P)^2 + 0.4024P]$ where $P = (F_o^2 + 2F_c^2)/3$
1392 reflections	$(\Delta/\sigma)_{max} = 0.001$
86 parameters	$\Delta\rangle_{max} = 0.29$ e Å^{-3}
1 restraint	$\Delta\rangle_{min} = -0.36$ e Å^{-3}

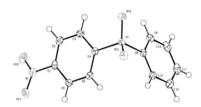

1-nitro-4-(phenylsulfonyl)benzene (172) SB1217_HY

Crystal data

C$_{12}$H$_9$NO$_4$S	D_x = 1.526 Mg m^{-3}
M_r = 263.26	Cu $K\alpha$ radiation, λ = 1.54178 Å
Orthorhombic, $Pna2_1$ (no.33)	Cell parameters from 9855 reflections
a = 7.1532 (2) Å	θ = 4.1–72.0°
b = 14.8966 (4) Å	μ = 2.60 mm^{-1}
c = 10.7512 (3) Å	T = 123 K
V = 1145.63 (5) Å^3	Blocks, colorless
Z = 4	0.20 × 0.12 × 0.04 mm
$F(000)$ = 544	

Data collection

Bruker D8 VENTURE diffractometer with PhotonII CPAD detector	2247 reflections with $I > 2\sigma(I)$
Radiation source: INCOATEC microfocus sealed tube	R_{int} = 0.021
rotation in ϕ and ω, 1°, shutterless scans	θ_{max} = 72.0°, θ_{min} = 5.9°
Absorption correction: multi-scan *SADABS* V2014/5 (Bruker AXS Inc.)	h = -8→8
T_{min} = 0.727, T_{max} = 0.841	k = -18→18
14242 measured reflections	l = -13→13
2247 independent reflections	

Refinement

Refinement on F^2	Secondary atom site location: difference Fourier map
Least-squares matrix: full	Hydrogen site location: difference Fourier map
$R[F^2 > 2\sigma(F^2)]$ = 0.021	H-atom parameters constrained
$wR(F^2)$ = 0.055	$w = 1/[\sigma^2(F_o^2) + (0.0356P)^2 + 0.2182P]$ where $P = (F_o^2 + 2F_c^2)/3$
S = 1.08	$(\Delta/\sigma)_{max} < 0.001$
2247 reflections	$\Delta\rangle_{max}$ = 0.27 e Å^{-3}
163 parameters	$\Delta\rangle_{min}$ = -0.20 e Å^{-3}
1 restraint	Absolute structure: Flack x determined using 1056 quotients [(I+)-(I-)]/[(I+)+(I-)] (Parsons, Flack and Wagner, Acta Cryst. B69 (2013) 249-259).
Primary atom site location: dual	Absolute structure parameter: 0.006 (6)

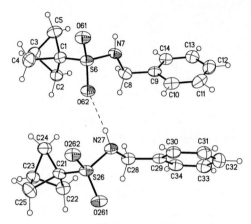

N-benzylbicyclo[1.1.1]pentane-1-sulfonamide (176b) SB1209_HY

Crystal data

$C_{12}H_{15}NO_2S$	$F(000) = 1008$
$M_r = 237.31$	$D_x = 1.341$ Mg m^{-3}
Monoclinic, $P2_1/n$ *(no.14)*	Cu $K\alpha$ radiation, $\lambda = 1.54178$ Å
$a = 5.7746$ (2) Å	Cell parameters from 9272 reflections
$b = 17.6889$ (5) Å	$\theta = 3.1–71.8°$
$c = 23.1791$ (7) Å	$\mu = 2.33$ mm^{-1}
$\beta = 96.793$ (2)°	$T = 123$ K
$V = 2351.04$ (13) Å^3	Plates, colorless
$Z = 8$	$0.24 \times 0.08 \times 0.02$ mm

Data collection

Bruker D8 VENTURE diffractometer with PhotonII CPAD detector	3714 reflections with $I > 2\sigma(I)$
Radiation source: INCOATEC microfocus sealed tube	$R_{int} = 0.062$
rotation in ϕ and ω, 1°, shutterless scans	$\theta_{max} = 72.0°, \theta_{min} = 3.2°$
Absorption correction: multi-scan *SADABS* V2014/5 (Bruker AXS Inc.)	$h = -6\rightarrow6$
$T_{min} = 0.726, T_{max} = 0.942$	$k = -21\rightarrow19$
23913 measured reflections	$l = -28\rightarrow28$
4548 independent reflections	

Refinement

Refinement on F^2	Primary atom site location: dual
Least-squares matrix: full	Secondary atom site location: difference Fourier map
$R[F^2 > 2\sigma(F^2)] = 0.061$	Hydrogen site location: difference Fourier map
$wR(F^2) = 0.148$	H atoms treated by a mixture of independent and constrained refinement
$S = 1.11$	$w = 1/[\sigma^2(F_o^2) + (0.0381P)^2 + 4.4665P]$ where $P = (F_o^2 + 2F_c^2)/3$
4548 reflections	$(\Delta/\sigma)_{max} = 0.001$
301 parameters	$\Delta\rangle_{max} = 0.52$ e Å^{-3}
4 restraints	$\Delta\rangle_{min} = -0.44$ e Å^{-3}

7.4.2 Crystallographic data solved by Dr. Olaf Fuhr

The following crystal structure (CCDC 1896794) was measured and solved by Dr. Olaf Fuhr at the Institute of Nanotechnology (KIT).

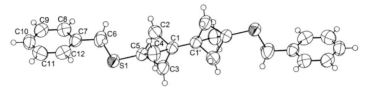

3,3'-bis(benzylthio)-1,1'-bi(bicyclo[1.1.1]pentane (148) GH021

Identification code	GH021
Empirical formula	$C_{24}H_{26}S_2$
Formula weight	378.57
Temperature/K	200.15
Crystal system	monoclinic
Space group	$P2_1/n$
a/Å	6.2148(9)
b/Å	11.2735(10)
c/Å	15.165(2)
α/°	90
β/°	99.751(12)
γ/°	90
Volume/Å^3	1047.1(2)
Z	2
ρ_{calc}g/cm^3	1.201
μ/mm^{-1}	1.506
F(000)	404.0
Crystal size/mm^3	0.23 × 0.04 × 0.03
Radiation	GaKα (λ = 1.34143)
2Θ range for data collection/°	19.89 to 115.066
Index ranges	-5 ≤ h ≤ 7, -14 ≤ k ≤ 13, -19 ≤ l ≤ 13
Reflections collected	7376
Independent reflections	2112 [R_{int} = 0.0339, R_{sigma} = 0.0365]
Indep. refl. with I>=2σ (I)	1433
Data/restraints/parameters	2112/0/118
Goodness-of-fit on F^2	1.057
Final R indexes [I>=2σ (I)]	R_1 = 0.0443, wR_2 = 0.1246
Final R indexes [all data]	R_1 = 0.0626, wR_2 = 0.1296
Largest diff. peak/hole / e Å^{-3}	0.20/-0.21

8   List of abbreviations

A

abs .... absolute

ADC .... antibody-drug conjugate

ATR .... attenuated total reflection

ATRA .... atom transfer radical addition

B

BCB .... bicyclo[1.1.0]butane

BCO .... bicyclo[2.2.2]octane

BCP .... bicyclo[1.1.1]pentane

BDE .... bond dissociation energy

BTK .... Bruton's tyrosine kinase

C

calc .... calculated

CDK .... cyclic-dependent kinase

C_q .... quaternary carbon

D

DEPT ...... distortionless enhancement by polarization
      transfer

DFT ....density functional theory

DMEDA .... N,N′-dimethylethylenediamine

DMG .... direct metalation group

DMSO .... dimethyl sulfoxide

dpm .... dipivaloylmethanat

dppe diphosphine 1,2-bis(diphenylphosphino)ethane

dppf .... 1,1′-bis(diphenylphosphino)ferrocene

E

ED_{50} .... effective dose

EDG ....electron-donating group

EI-MS ....electron-ionization-mass spectrometry

equiv ....equivalents

ESI-MS ...... eletrospray ionization-mass spectrometry

EWG .... electron-withdrawing group

F

Fab..................................... fragment antigen binding

FAB-MS.. fast atom bombarment-mass spectrometry

FDA .... Food and Drug Administration

G

GC-MS .... gas chromatography-mass spectrometry

H

HMBC ...heteronuclear multiple quantum correlation

HPLC .... high-performance liquid chromatography

HR-MS .... high resolution-mass spectrometry

HSQC .... heteronuclear single quantum coherence

I

IR .... infrared spectroscopy

IUPAC ....International Union of Pure and Applied
      Chemistry

L

$LpPLA_2$ .... lipoprotein-associated phospholipase A2

M

mCPBA............................... m-chloroperbenzoic acid

mGluR................... metabotropic glutamate receptor

MOP .... multi-objective problem

N

NHS..N-hydroxysuccinimidyl

NMR ....nuclear magnetic resonance

NRH .... non-conjugated rigid hydrocarbon

P

p.a. .... *pro analysis*

PABA ....*para*-aminobenzoic acid

pK_a....................................... acid dissociation constant

Q

quant.. quantitative

S

SAR..............................structure-activity relationship

SFV .... Semliki Forest virus

T

TEMPO .... (2,2,6,6-tetramethylpiperidin-1-yl)oxyl

Tf .... triflate

TFA .... trifluoroacetic acid

TFDA...................... trimethylsilyl 2 fluorosulfonyl-2,2

difluoroacetate

THF... tetrahydrofuran

TLC .... thin layer chromatography

TTMSS ....tris(trimethylsilyl)silane

U

UV .... ultraviolet

9 References

[1]   Bohacek, R. S., McMartin, C., Guida, W. C., *Med. Res. Rev.* **1996**, *16*, 3–50.

[2]   Dobson, C. M., *Nature* **2004**, *432*, 824–828.

[3]   Opassi, G., Gesù, A., Massarotti, A., *Drug Discov. Today* **2018**, *23*, 565–574.

[4]   Triggle, D. J., *Biochem. Pharmacol.* **2009**, *78*, 217–223.

[5]   Verdine, G. L., *Nature* **1996**, *384*, 11–13.

[6]   Schreiber, S. L., *Science* **2000**, *287*, 1964–1969.

[7]   Hajduk, P. J., Greer, J., *Nat. Rev. Drug Discov.* **2007**, *6*, 211–219.

[8]   Brown, D. G., Gagnon, M. M., Boström, J., *J. Med. Chem.* **2015**, *58*, 2390–2405.

[9]   Brown, D. G., Boström, J., *J. Med. Chem.* **2016**, *59*, 4443–4458.

[10]   Boström, J., Brown, D. G., Young, R. J., Keserü, G. M., *Nat. Rev. Drug Discov.* **2018**, *17*, 709–727.

[11]   Nicolaou, C. A., Brown, N., *Drug Discov. Today Technol.* **2013**, *10*, e427–e435.

[12]   Brown, N., in *Bioisosteres in Medicinal Chemistry* (Ed.: H. K. R. Mannhold, G. Folkers, N. Brown), Wiley Online Library, Weinheim, Germany, **2012**, pp. 1–14.

[13]   Martin, Y. C., *J. Med. Chem.* **1981**, *24*, 229–237.

[14]   Abraham, D. J., *Burger's medicinal chemistry and drug discovery, Vol. 1*, 6 ed., Wiley, Hoboken, N.J., **2003**.

[15]   Patani, G. A., LaVoie, E. J., *Chem. Rev.* **1996**, *96*, 3147–3176.

[16]   Grimm, H. G., *Z. Electrochem.* **1925**, *31*, 474–480.

[17]   Grimm, H. G., *Naturwissenschaften* **1929**, *17*, 557–564.

[18]   Erlenmeyer, H., Leo, M., *Helv. Chim. Acta* **1932**, *15*, 1171–1186.

[19]   Foye, W. O., *Foye's principles of medicinal chemistry*, Lippincott Williams & Wilkins, **2008**.

[20]   Rosenblum, S. B., Huynh, T., Afonso, A., Davis, H. R., Yumibe, N., Clader, J. W., Burnett, D. A., *J. Med. Chem.* **1998**, *41*, 973–980.

[21]   Bonnet, P. A., Robins, R. K., *J. Med. Chem.* **1993**, *36*, 635–653.

[22]   Sauerberg, P., Chen, J., WoldeMussie, E., Rapoport, H., *J. Med. Chem.* **1989**, *32*, 1322–1326.

[23]   Dunn, D. L., Scott, B. S., Dorsey, E. D., *J. Pharm. Sci.* **1981**, *70*, 446–449.

[24]   Larsen, A. A., Lish, P. M., *Nature* **1964**, *203*, 1283–1284.

[25]   Domagk, G., *Dtsch. Med. Wochenschr.* **1935**, *61*, 250–253.

[26]   Třefouël, J., Třefouël, J. M., Nitti, F., Bovet, D., *Compt. Rend. Soc. Bio.* **1935**, *120*, 756–758.

[27]   Then, R., Angehrn, P., *Microbiology* **1973**, *76*, 255–263.

[28]   Dodds, E. C., Goldberg, L., Lawson, W., Robinson, R., *Nature* **1938**, *141*, 247–248.

[29]   Walton, E., Brownlee, G., *Nature* **1943**, *151*, 305–306.

[30]   Blanchard, E., Stuart, A., Tallman, R., *Endocrinology* **1943**, *32*, 307–309.

[31]   Baker, B. R., *J. Am. Chem. Soc.* **1943**, *65*, 1572–1579.

[32]   Locke, G. M., Bernhard, S. S. R., Senge, M. O., *Chem. Eur. J.* **2019**, *25*, 4590–4647.

[33]   Lovering, F., Bikker, J., Humblet, C., *J. Med. Chem.* **2009**, *52*, 6752–6756.

[34]   Lovering, F., *MedChemComm* **2013**, *4*, 515–519.

[35]   Mykhailiuk, P. K., *Org. Biomol. Chem.* **2019**, *17*, 2839–2849.

[36]   Westphal, M. V., Wolfstädter, B. T., Plancher, J.-M., Gatfield, J., Carreira, E. M., *ChemMedChem* **2015**, *10*, 461–469.

[37]   Pellicciari, R., Costantino, G., Giovagnoni, E., Mattoli, L., Brabet, I., Pin, J.-P., *Bioorg. Med. Chem. Lett.* **1998**, *8*, 1569–1574.

[38]   Pellicciari, R., Raimondo, M., Marinozzi, M., Natalini, B., Costantino, G., Thomsen, C., *J. Med. Chem.* **1996**, *39*, 2874–2876.

[39]   Stepan, A. F., Subramanyam, C., Efremov, I. V., Dutra, J. K., O'Sullivan, T. J., DiRico, K. J., McDonald, W. S., Won, A., Dorff, P. H., Nolan, C. E., Becker, S. L., Pustilnik, L. R., Riddell, D. R., Kauffman, G. W., Kormos, B. L., Zhang, L., Lu, Y., Capetta, S. H., Green, M. E., Karki, K., Sibley, E., Atchison, K. P., Hallgren, A. J., Oborski, C. E., Robshaw, A. E., Sneed, B., O'Donnell, C. J., *J. Med. Chem.* **2012**, *55*, 3414–3424.

[40]   Gao, X., Wang, J., Liu, J., Guiadeen, D., Krikorian, A., Boga, S. B., Alhassan, A.-B., Selyutin, O., Yu, W., Yu, Y., Anand, R., Liu, S., Yang, C., Wu, H., Cai, J., Cooper, A., Zhu, H., Maloney, K., Gao, Y.-D., Fischmann, T. O., Presland, J., Mansueto, M., Xu, Z., Leccese, E., Zhang-Hoover, J., Knemeyer, I., Garlisi, C. G., Bays, N., Stivers, P., Brandish, P. E., Hicks, A., Kim, R., Kozlowski, J. A., *Bioorg. Med. Chem. Lett.* **2017**, *27*, 1471–1477.

[41]   Coric, V., van Dyck, C. H., Salloway, S., Andreasen, N., Brody, M., Richter, R. W., Soininen, H., Thein, S., Shiovitz, T., Pilcher, G., Colby, S., Rollin, L., Dockens, R., Pachai, C., Portelius, E., Andreasson, U., Blennow, K., Soares, H., Albright, C., Feldman, H. H., Berman, R. M., *Arch. Neurol.* **2012**, *69*, 1430–1440.

[42]   Hermanson, G. T., *Bioconjugate techniques*, Academic press, **2013**.

[43]   Stephanopoulos, N., Francis, M. B., *Nat. Chem. Biol.* **2011**, *7*, 876–884.

[44]   Begley, T. P., Tilley, S. D., Joshi, N. S., Francis, M. B., in *Wiley Encyclopedia of Chemical Biology*, **2008**, pp. 1-16.

[45]   Fasman, G. D., *Prediction of protein structure and the principles of protein conformation*, Springer Science & Business Media, **2012**.

[46]   Shaunak, S., Godwin, A., Choi, J.-W., Balan, S., Pedone, E., Vijayarangam, D., Heidelberger, S., Teo, I., Zloh, M., Brocchini, S., *Nat. Chem. Biol.* **2006**, *2*, 312–313.

[47]   Niida, A., Sasaki, S., Yonemori, K., Sameshima, T., Yaguchi, M., Asami, T., Sakamoto, K., Kamaura, M., *Bioorg. Med. Chem. Lett.* **2017**, *27*, 2757–2761.

[48]   Chari, R. V. J., Miller, M. L., Widdison, W. C., *Angew. Chem. Int. Ed.* **2014**, *53*, 3796–3827.

[49]   Griebenow, N., Dilmaç, A. M., Greven, S., Bräse, S., *Bioconjugate Chem.* **2016**, *27*, 911–917.

[50]   Dilmaç, A. M., Spuling, E., de Meijere, A., Bräse, S., *Angew. Chem. Int. Ed.* **2017**, *56*, 5684–5718.

[51]   Wiberg, K. B., *J. Am. Chem. Soc.* **1983**, *105*, 1227–1233.

[52]   Wiberg, K. B., Dailey, W. P., Walker, F. H., Waddell, S. T., Crocker, L. S., Newton, M., *J. Am. Chem. Soc.* **1985**, *107*, 7247–7257.

[53]   Wiberg, K. B., *Chem. Rev.* **1989**, *89*, 975–983.

[54]   Zalkow, L. H., Harris, R. N., Van Derveer, D., *J. Chem. Soc., Chem. Commun.* **1978**, *1978*, 420–421.

[55]   Nied, D., Klopper, W., Breher, F., *Angew. Chem. Int. Ed.* **2009**, *48*, 1411–1416.

[56]   Altman, J., Babad, E., Itzchaki, J., Ginsburg, D., *Tetrahedron* **1966**, *22*, 279–304.

[57]   Wiberg, K. B., Walker, F. H., *J. Am. Chem. Soc.* **1982**, *104*, 5239–5240.

[58]   Feller, D., Davidson, E. R., *J. Am. Chem. Soc.* **1987**, *109*, 4133–4139.

[59]   Jackson, J. E., Allen, L. C., *J. Am. Chem. Soc.* **1984**, *106*, 591–599.

[60]   Wiberg, K. B., *Angew. Chem. Int. Ed.* **1986**, *25*, 312–322.

[61]   Wu, W., Gu, J., Song, J., Shaik, S., Hiberty, P. C., *Angew. Chem. Int. Ed.* **2009**, *48*, 1407–1410.

[62]   Laplaza, R., Contreras-Garcia, J., Fuster, F., Volatron, F., Chaquin, P., *Chem. Eur. J.*, doi:10.1002/chem.201904910.

[63]   Semmler, K., Szeimies, G., Belzner, J., *J. Am. Chem. Soc.* **1985**, *107*, 6410–6411.

[64]   Belzner, J., Gareiß, B., Polborn, K., Schmid, W., Semmler, K., Szeimies, G., *Chem. Ber.* **1989**, *122*, 1509–1529.

[65]   Lynch, K. M., Dailey, W. P., *J. Org. Chem.* **1995**, *60*, 4666–4668.

[66] Gianatassio, R., Lopchuk, J. M., Wang, J., Pan, C.-M., Malins, L. R., Prieto, L., Brandt, T. A., Collins, M. R., Gallego, G. M., Sach, N. W., Spangler, J. E., Zhu, H., Zhu, J., Baran, P. S., *Science* **2016**, *351*, 241–246.

[67] Alber, F., Szeimies, G., *Chem. Ber.* **1992**, *125*, 757–758.

[68] Mondanaro, K. R., Dailey, W. P., *Org. Synth.* **1998**, *75*, 98–105.

[69] Belzner, J., Szeimies, G., *Tetrahedron Lett.* **1986**, *27*, 5839–5842.

[70] Levin, M. D., Kaszynski, P., Michl, J., *Chem. Rev.* **2000**, *100*, 169–234.

[71] Kanazawa, J., Uchiyama, M., *Synlett* **2019**, *30*, 1–11.

[72] Rehm, J. D. D., Ziemer, B., Szeimies, G., *Eur. J. Org. Chem.* **1999**, *1999*, 2079–2085.

[73] Wiberg, K. B., Waddell, S. T., *J. Am. Chem. Soc.* **1990**, *112*, 2194–2216.

[74] Wiberg, K. B., McMurdie, N., *J. Am. Chem. Soc.* **1991**, *113*, 8995–8996.

[75] Wiberg, K. B., McMurdie, N., *J. Am. Chem. Soc.* **1994**, *116*, 11990–11998.

[76] Münch, S. W., PhD thesis, Karlsruhe Institute of Technology (KIT) **2018**.

[77] Kaszynski, P., Michl, J., *J. Org. Chem.* **1988**, *53*, 4593–4594.

[78] Messner, M., Kozhushkov, Sergei I., de Meijere, A., *Eur. J. Org. Chem.* **2000**, *2000*, 1137–1155.

[79] Rehm, J. D. D., Ziemer, B., Szeimies, G., *Eur. J. Org. Chem.* **2001**, *2001*, 1049–1052.

[80] Makarov, I. S., Brocklehurst, C. E., Karaghiosoff, K., Koch, G., Knochel, P., *Angew. Chem. Int. Ed.* **2017**, *56*, 12774–12777.

[81] Caputo, D. F. J., Arroniz, C., Dürr, A. B., Mousseau, J. J., Stepan, A. F., Mansfield, S. J., Anderson, E. A., *Chem. Sci.* **2018**, *9*, 5295–5300.

[82] Nugent, J., Arroniz, C., Shire, B. R., Sterling, A. J., Pickford, H. D., Wong, M. L. J., Mansfield, S. J., Caputo, D. F. J., Owen, B., Mousseau, J. J., Duarte, F., Anderson, E. A., *ACS Catal.* **2019**, *9*, 9568–9574.

[83] Trongsiriwat, N., Pu, Y., Nieves-Quinones, Y., Shelp, R. A., Kozlowski, M. C., Walsh, P. J., *Angew. Chem. Int. Ed.* **2019**, *58*, 13416–13420.

[84] Bunker, K. D., Sach, N. W., Huang, Q., Richardson, P. F., *Org. Lett.* **2011**, *13*, 4746–4748.

[85] Lopchuk, J. M., Fjelbye, K., Kawamata, Y., Malins, L. R., Pan, C.-M., Gianatassio, R., Wang, J., Prieto, L., Bradow, J., Brandt, T. A., Collins, M. R., Elleraas, J., Ewanicki, J., Farrell, W., Fadeyi, O. O., Gallego, G. M., Mousseau, J. J., Oliver, R., Sach, N. W., Smith, J. K., Spangler, J. E., Zhu, H., Zhu, J., Baran, P. S., *J. Am. Chem. Soc.* **2017**, *139*, 3209–3226.

[86] Kanazawa, J., Maeda, K., Uchiyama, M., *J. Am. Chem. Soc.* **2017**, *139*, 17791–17794.

[87] Hughes, J. M. E., Scarlata, D. A., Chen, A. C. Y., Burch, J. D., Gleason, J. L., *Org. Lett.* **2019**, *21*, 6800–6804.

[88] Rout, S. K., Marghem, G., Lan, J., Leyssens, T., Riant, O., *Chem. Commun.* **2019**, *55*, 14976–14979.

[89] Quiclet-Sire, B., Zard, S. Z., *Chem. Eur. J.* **2006**, *12*, 6002–6016.

[90] Applequist, D. E., Wheeler, J. W., *Tetrahedron Lett.* **1977**, *18*, 3411–3412.

[91] Measom, N. D., Down, K. D., Hirst, D. J., Jamieson, C., Manas, E. S., Patel, V. K., Somers, D. O., *ACS Med. Chem. Lett.* **2017**, *8*, 43–48.

[92] Fieser, L. F., Sachs, D. H., *J. Org. Chem.* **1964**, *29*, 1113–1115.

[93] Ma, X., Sloman, D. L., Han, Y., Bennett, D. J., *Org. Lett.* **2019**, *21*, 7199–7203.

[94] Bychek, R. M., Hutskalova, V., Bas, Y. P., Zaporozhets, O. A., Zozulya, S., Levterov, V. V., Mykhailiuk, P. K., *J. Org. Chem.* **2019**, *84*, 15106–15117.

[95] Kenndoff, M., Singer, A., Szeimies, G., *J. Prakt. Chem.* **1997**, *339*, 217–232.

[96] Bär, R. M., Master thesis, Karlsruhe Institute of Technology (KIT) **2016**.

[97] "METHYL BROMIDE", ILO International Chemical Safety Cards (ICSC), accessed 07.11.2019, http://www.ilo.org/dyn/icsc/showcard.display?p_version=2&p_card_id=0109.

[98]   "BROMOBENZENE", ILO International Chemical Safety Cards (ICSC), accessed
        07.11.2019,
        http://www.ilo.org/dyn/icsc/showcard.display?p_version=2&p_card_id=1016.
[99]   Merchant, R. R., Edwards, J. T., Qin, T., Kruszyk, M. M., Bi, C., Che, G., Bao, D.-H.,
        Qiao, W., Sun, L., Collins, M. R., Fadeyi, O. O., Gallego, G. M., Mousseau, J. J., Nuhant,
        P., Baran, P. S., *Science* **2018**, *360*, 75–80.
[100]   Marinozzi, M., Fulco, M. C., Rizzo, R., Pellicciari, R., *Synlett* **2004**, *2004*, 1027–1028.
[101]   Boger, D. L., Mathvink, R. J., *J. Org. Chem.* **1992**, *57*, 1429–1443.
[102]   Della, E., Tsanaktsidis, J., *Austr. J. Chem.* **1986**, *39*, 2061–2066.
[103]   Della, E., Taylor, D., *Austr. J. Chem.* **1990**, *43*, 945–948.
[104]   Della, E., Taylor, D., *Austr. J. Chem.* **1991**, *44*, 881–885.
[105]   Nauser, T., Dockheer, S., Kissner, R., Koppenol, W. H., *Biochemistry* **2006**, *45*, 6038–
        6043.
[106]   Chan, W., White, P., *Fmoc solid phase peptide synthesis: a practical approach, Vol. 222*,
        OUP Oxford, **1999**.
[107]   "bicyclo[1.1.1]pentane-1-thiol",   Enanime   Ltd.,   accessed   12.11.2019,
        https://www.enaminestore.com/catalog.
[108]   McGarry, P., Johnston, L., Scaiano, J., *J. Org. Chem.* **1989**, *54*, 6133–6135.
[109]   Wiberg, K. B., Waddell, S. T., Laidig, K., *Tetrahedron Lett.* **1986**, *27*, 1553–1556.
[110]   Wiberg, K. B., Waddell, S. T., *Tetrahedron Lett.* **1988**, *29*, 289–292.
[111]   Bunz, U., Polborn, K., Wagner, H.-U., Szeimies, G., *Chem. Ber.* **1988**, *121*, 1785–1790.
[112]   Friedli, A. C., Kaszynski, P., Michl, J., *Tetrahedron Lett.* **1989**, *30*, 455–458.
[113]   Kaszynski, P., Friedli, A. C., Michl, J., *J. Am. Chem. Soc.* **1992**, *114*, 601–620.
[114]   Kaszynski, P., Michl, J., *J. Am. Chem. Soc.* **1988**, *110*, 5225–5226.
[115]   Bent, H. A., *Chem. Rev.* **1961**, *61*, 275–311.
[116]   Murthy, G. S., Hassenruck, K., Lynch, V. M., Michl, J., *J. Am. Chem. Soc.* **1989**, *111*,
        7262–7264.
[117]   Yang, Y.-M., Yu, H.-Z., Sun, X.-H., Dang, Z.-M., *J. Phys. Org. Chem.* **2016**, *29*, 6–13.
[118]   Langer, L., Master thesis, Karlsruhe Institute of Technology (KIT) **2019**.
[119]   Feng, M., Tang, B., Liang, S. H., Jiang, X., *Curr. Top. Med. Chem.* **2016**, *16*, 1200–1216.
[120]   Brunel, J.-M., Diter, P., Duetsch, M., Kagan, H. B., *J. Org. Chem.* **1995**, *60*, 8086–8088.
[121]   Yamaguchi, T., Matsumoto, K., Saito, B., Katsuki, T., *Angew. Chem. Int. Ed.* **2007**, *46*,
        4729–4731.
[122]   Snieckus, V., *Chem. Rev.* **1990**, *90*, 879–933.
[123]   Iwao, M., Iihama, T., Mahalanabis, K., Perrier, H., Snieckus, V., *J. Org. Chem.* **1989**, *54*,
        24–26.
[124]   Maksić, Z. B., Eckert-Maksić, M., *Tetrahedron* **1969**, *25*, 5113–5114.
[125]   Della, E. W., Cotsaris, E., Hine, P., Pigou.P.E, *Australian Journal of Chemistry* **1981**, *34*,
        913–916.
[126]   Rhodes, C. J., Walton, J. C., Della, E. W., *J. Chem. Soc., Perkin Trans. 2* **1993**, *1993*,
        2125–2128.
[127]   Muller, N., Pritchard, D. E., *J. Chem. Phys.* **1959**, *31*, 768–771.
[128]   Muller, N., Pritchard, D. E., *J. Chem. Phys.* **1959**, *31*, 1471–1476.
[129]   Garlets, Z. J., Sanders, J. N., Malik, H., Gampe, C., Houk, K. N., Davies, H. M. L., *Nat.
        Catal.* **2020**, doi:10.1038/s41929-019-0417-1.
[130]   Lücking, U., *Angew. Chem. Int. Ed.* **2013**, *52*, 9399–9408.
[131]   Lücking, U., *Org. Chem. Front.* **2019**, *6*, 1319–1324.
[132]   Frings, M., Bolm, C., Blum, A., Gnamm, C., *Eur. J. Med. Chem.* **2017**, *126*, 225–245.
[133]   Reck, M., Horn, L., Novello, S., Barlesi, F., Albert, I., Juhasz, E., Chung, J., Fritsch, A.,
        Drews, U., Rutstein, M., Wagner, A., Govindan, R., *Ann. Oncol.* **2016**, *27*.

[134] Cho, B. C., Dy, G. K., Govindan, R., Kim, D.-W., Pennell, N. A., Zalcman, G., Besse, B., Kim, J.-H., Koca, G., Rajagopalan, P., Langer, S., Ocker, M., Nogai, H., Barlesi, F., *Lung Cancer* **2018**, *123*, 14–21.

[135] Nishimura, N., Norman, M. H., Liu, L., Yang, K. C., Ashton, K. S., Bartberger, M. D., Chmait, S., Chen, J., Cupples, R., Fotsch, C., Helmering, J., Jordan, S. R., Kunz, R. K., Pennington, L. D., Poon, S. F., Siegmund, A., Sivits, G., Lloyd, D. J., Hale, C., St. Jean, D. J., *J. Med. Chem.* **2014**, *57*, 3094–3116.

[136] Oost, T., Fiegen, D., Gnamm, C., Handschuh, S., Peters, S., Roth, G. J., *Substituted 4-pyridones and their use as inhibitors of neutrophil elastase activity*, **2015**, U.S. Patent No. 9,102,624.

[137] Bizet, V., Hendriks, C. M. M., Bolm, C., *Chem. Soc. Rev.* **2015**, *44*, 3378–3390.

[138] Xie, Y., Zhou, B., Zhou, S., Zhou, S., Wei, W., Liu, J., Zhan, Y., Cheng, D., Chen, M., Li, Y., Wang, B., Xue, X.-s., Li, Z., *ChemistrySelect* **2017**, *2*, 1620–1624.

[139] Zenzola, M., Doran, R., Degennaro, L., Luisi, R., Bull, J. A., *Angew. Chem. Int. Ed.* **2016**, *55*, 7203–7207.

[140] Hosseinian, A., Zare Fekri, L., Monfared, A., Vessally, E., Nikpassand, M., *J. Sulfur Chem.* **2018**, *39*, 674–698.

[141] Correa, A., Bolm, C., *Adv. Synth. Catal.* **2007**, *349*, 2673–2676.

[142] Kondo, M., Kanazawa, J., Ichikawa, T., Shimokawa, T., Nagashima, Y., Miyamoto, K., Uchiyama, M., *Angew. Chem. Int. Ed.* **2020**, *59*, 1970–1974.

[143] Scott, K. A., Njardarson, J. T., *Top. Curr. Chem.* **2018**, *376*, 1–34.

[144] Vogel, A. I., Tatchell, A., Furnis, B., Hannaford, A., Smith, P., *Vogel's textbook of practical organic chemistry*, John Wiley & Sons, Inc., New York, **1989**.

[145] Baskin, J. M., Wang, Z., *Tetrahedron Lett.* **2002**, *43*, 8479–8483.

[146] Sato, K., Hyodo, M., Aoki, M., Zheng, X.-Q., Noyori, R., *Tetrahedron* **2001**, *57*, 2469–2476.

[147] Kaiser, D., Klose, I., Oost, R., Neuhaus, J., Maulide, N., *Chem. Rev.* **2019**, *119*, 8701–8780.

[148] Pan, X., Gao, J., Liu, J., Lai, J., Jiang, H., Yuan, G., *Green Chem.* **2015**, *17*, 1400–1403.

[149] Ammazzalorso, A., De Filippis, B., Giampietro, L., Amoroso, R., *Chem. Biol. Drug Des.* **2017**, *90*, 1094–1105.

[150] Michaudel, Q., Ishihara, Y., Baran, P. S., *Acc. Chem. Res.* **2015**, *48*, 712–721.

[151] Yan, M., Lo, J. C., Edwards, J. T., Baran, P. S., *J. Am. Chem. Soc.* **2016**, *138*, 12692–12714.

[152] Gianatassio, R., Kawamura, S., Eprile, C. L., Foo, K., Ge, J., Burns, A. C., Collins, M. R., Baran, P. S., *Angew. Chem. Int. Ed.* **2014**, *53*, 9851–9855.

[153] Knauber, T., Chandrasekaran, R., Tucker, J. W., Chen, J. M., Reese, M., Rankic, D. A., Sach, N., Helal, C., *Org. Lett.* **2017**, *19*, 6566–6569.

[154] Woolven, H., González-Rodríguez, C., Marco, I., Thompson, A. L., Willis, M. C., *Org. Lett.* **2011**, *13*, 4876–4878.

[155] Lenstra, D. C., Vedovato, V., Ferrer Flegeau, E., Maydom, J., Willis, M. C., *Org. Lett.* **2016**, *18*, 2086–2089.

[156] Gottlieb, H. E., Kotlyar, V., Nudelman, A., *J. Org. Chem.* **1997**, *62*, 7512–7515.

[157] Sheldrick, G. M., *Acta Cryst.* **2015**, *A71*, 3–8.

[158] Sheldrick, G. M., *Acta Cryst.* **2015**, *C71*, 3–8.

[159] Dolomanov, O. V., Bourhis, L. J., Gildea, R. J., Howard, J. A. K., Puschmann, H., *J. Appl. Crystallogr.* **2009**, *42*, 339–341.

[160] Sheldrick, G. M., *Acta Cryst.* **2008**, *A64*, 112–122.

[161] Puig-de-la-Bellacasa, R., Giménez, L., Pettersson, S., Pascual, R., Gonzalo, E., Esté, J. A., Clotet, B., Borrell, J. I., Teixidó, J., *Eur. J. Med. Chem.* **2012**, *54*, 159–174.

[162] Chen, D., Shi, G., Jiang, H., Zhang, Y., Zhang, Y., *Org. Lett.* **2016**, *18*, 2130–2133.

[163] Nishino, K., Ogiwara, Y., Sakai, N., *Eur. J. Org. Chem.* **2017**, *2017*, 5892–5895.
[164] Cossu, S., Delogu, G., Fabbri, D., Maglioli, P., *Org. Prep. Proced. Int.* **1991**, *23*, 455–457.
[165] Bian, M., Xu, F., Ma, C., *Synthesis* **2007**, *2007*, 2951–2956.

10 Appendix

10.1 Curriculum Vitae

Personal information

Name	Robin Maximilian Bär
Born	26.09.1990 in Pforzheim
Address	Augartenstr. 15, 76137 Karlsruhe
Mobile	+49 176 31204356
Email	robinbaer@gmx.de

Education

01/2017 – 02/2020 **Ph.D.**, Institute of Organic Chemistry at the Karlsruhe Institute of Technology (KIT), group of Prof. Dr. Stefan Bräse

Titel of the PhD Thesis

Synthesis and modification of bicyclo[1.1.1]pentyl sulfides

10/2014 – 09/2016 **M.Sc. (1.0)**, Chemical Biology at the Karlsruhe Institute of Technology (KIT), Focus on Chemical Biology and Organic Chemistry

Titel of the Master Thesis

Synthesis of novel [1.1.1]propellanes and bicyclo[1.1.1]pentanes, mark 1.0

10/2011 – 09/2014 **B.Sc. (1.8)**, Chemical Biology at the Karlsruhe Institute of Technology (KIT), Focus on Biochemistry and Chemical Biology

Titel of the Bachelor Thesis

Full-cell measurements of microbial fuel cells with Shewanella oneidensis, mark 1.7

Experience/Internships

04/2019 – 09/2019 **External stay**, School of Chemistry at the University of Bristol, UK, group of Prof. Varinder K. Aggarwal

09/2018 – 11/2018 **Internship**, Boehringer Ingelheim Pharma GmbH & Co. KG, Biberach an der Riß

08/2016 – 12/2016 **Research Assistant**, Department of Applied Electrochemistry at the Fraunhofer Institute of Chemical Technology ICT, Pfinztal

01/2016 – 03/2016 **Internship**, Bayer Pharma AG, Wuppertal

06/2013 – 12/2015 **Research Assistant**, Department of Applied Electrochemistry at the Fraunhofer Institute of Chemical Technology ICT, Pfinztal

10.2 List of publications

Original Publications (peer-reviewed)

10. **Sodium Bicyclo[1.1.1]pentanesulfinate, a Bench-stable Precursor for Bicyclo[1.1.1]pentylsulfones and Bicyclo[1.1.1]pentanesulfonamides**

    R. M. Bär, Patrick J. Gross, M. Nieger, S. Bräse

    Chemistry - A European Journal, **2020**, *26*, 4242–4245, DOI:10.1002/chem.202000097

9. **Bicyclo[1.1.1]pentyl Sulfoximines: Synthesis and Functionalizations**

    R. M. Bär, L. Langer, M. Nieger, S. Bräse

    Advanced Synthesis & Catalysis, **2020**, *362*, 1356–1361, DOI:10.1002/adsc.201901453

8. **Photoinduced Deoxygenative Borylations of Aliphatic Alcohols**

    J. Wu, R. M. Bär, L. Guo, A. Noble, V. K. Aggarwal

    Angewandte Chemie International Edition, **2019**, *58*, 18830–18834, DOI:10.1002/anie.201910051

7. **Insertion of [1.1.1]propellane into aromatic disulfides**

    R. M. Bär, G. Heinrich, M. Nieger, O. Fuhr, S. Bräse

    Beilstein Journal of Organic Chemistry, **2019**, *15*, 1172–1180, DOI:10.3762/bjoc.15.114

6. **Alkyl and aryl thiol addition to [1.1.1]propellane: scope and limitations of a fast conjugation reaction**

    R. M. Bär, S. Kirschner, M. Nieger, S. Bräse

    Chemistry - A European Journal, **2018**, *24*, 1373–1382, DOI:10.1002/chem.201704105

5. **Facile fabrication of robust superhydrophobic surfaces: Comparative investigation**

    R. M. Bär, S. Widmaier, P. A. Levkin

    RSC Advances, **2016**, *6*, 98257–98266, DOI:10.1039/C6RA22336B

4. **On the Design of a Comb-Shaped, Poly (phenylene oxide)-Based Anodic Binder for Anion-Exchange Membrane Direct Methanol Fuel Cell (AEM-DMFC)**

    T. Jurzinsky, R. Bär, N. Heppe, M. Kübler, F. Jung, C. Cremers, J. Tübke

    ECS Transactions, **2016**, *75*, 1041–1054, DOI:10.1149/07514.1041ecst

3. **Methanol Oxidation on Ru- and Ni-Modified Pd-Electrocatalysts in Alkaline Media: A Comparative Differential Electrochemical Mass Spectrometry Study**

    T. Jurzinsky, B. Kintzel, R. Bär, C. Cremers, K. Pinkwart, J. Tübke

    ECS Transactions, **2016**, *75*, 983–995, DOI:10.1149/07514.0983ecst

2.   **Methanol oxidation on PdRh/C electrocatalyst in alkaline media: Temperature and methanol concentration dependencies**

T. Jurzinsky, B. Kintzel, R. Bär, C. Cremers, J. Tübke

Journal of Electroanalytical Chemistry, **2016**, *776*, 49–52,

DOI:10.1016/j.jelechem.2016.06.038

1.   **Highly active carbon supported palladium-rhodium Pd X Rh/C catalysts for methanol electrooxidation in alkaline media and their performance in anion exchange direct methanol fuel cells (AEM-DMFCs)**

T. Jurzinsky, R. Bär, C. Cremers, J. Tübke, P. Elsner

Electrochimica Acta, **2015**, *176*, 1191–1201, DOI:10.1016/j.electacta.2015.07.176

Reviews (peer-reviewed)

1.   **Occurence, synthesis and applications of [3.3.3]propellanes**

T. Wezeman, A. M. Dilmaç, R. M. Bär, S. Bräse

Natural Product Reports, **2020**, *37*, 224–245, DOI:10.1039/C8NP00086G

Poster presentations

2.   **Sulphur-containing bicyclo[1.1.1]pentanes from [1.1.1]propellane**

R. M. Bär, S. Bräse

16th Belgian Organic Synthesis Symposium, Brussels, Belgium, July 2018

1.   **Novel bicyclo[1.1.1]pentane (BPC) building blocks by thiol addition to [1.1.1]propellane**

R. M. Bär, S. Bräse

1st Alpine Winter Conference on Medicinal and Synthetic Chemistry, St. Anton, Austria, January 2018

10.3 Acknowledgements

I want to thank **Prof. Dr. Stefan Bräse** for being my supervisor and giving me the opportunity to obtain my PhD in such a great environment. From the first projects during my master thesis to the very last one in this work, you gave me freedom to explore and advice whenever I needed it. Balancing trust and control is a tough task, but you are mastering it.

Further I want to thank **Prof. Dr. Joachim Podlech**, who agreed to by my coreferent. I appreciate the interest in my research and the nice atmosphere during the defense.

For my stay at Boehringer Ingelheim I want to thank **Dr. Patrick Gross**. Those three months were a very efficient time of my PhD and I'm really satisfied with the project we worked on. Thank you for the countless discussions about chemistry and the insight into the work of a lab head you gave me. Of course, I also want to thank **Tobias Kramer** and **Heike Schültingkemper** for the great atmosphere in the lab and the helpful practical tips.

My external stay at the University of Bristol was a unique experience and I want to thank **Prof. Varinder Aggarwal** for having me in his group and for the incredible supervision. There are only few people with such a knowledge of details in chemical reactions and a never-ending curiosity. I'll always remember the time in your group. A special thanks to **Dr. Jingjing Wu**, **Dr. Lin Guo** and **Dr. Adam Noble** for the successful teamwork. And of course the rest of the Aggarwal group who made the six months feel like one. The funding by the **Karlsruhe House of Young Scientists (KHYS)** and the **CRC 1176** for my external stay is greatly acknowledged.

Thank you **Dr. Kevin Lam** for showing such great interest in the BCP compounds. I really enjoy discussing ideas with you and I'm sure we'll find a nice method within our project.

I want to thank **Prof. Dr. Matthias Olzmann** for discussions about kinetics and for a helpful model of the insertion of propellane into disulfides.

Also thank you **Dr. Antonia Stepan** for inspiring discussions about bioisosteres and future goals of our research. I hope that someday, we'll finally manage to work together on a project.

Thank you **Dr. Martin Nieger** and **Dr. Olaf Fuhr** for measuring and analyzing single-crystal X-ray diffractions. Thank you **Dr. Norbert Foitzik**, **Angelika Mösle**, **Lara Hirsch**, **Rieke Schulte**, **Dr. Andreas Rapp**, **Pia Lang**, **Tanja Ohmer-Scherrer**, **Lennart Oberle** and **Karolin Kohnle** for maintenance and operation of analytical devices. Thank you ComPlat-Team for taking care of many, many samples and providing data whenever needed.

A big thank you to the organizational talents **Christiane Lampert**, **Seline Samur**, **Janine Bolz** and **Cornelia Weber**.

During my PhD I was accompanied by students who had a great impact on my work and helped me to develop my leadership ability. Thank you **Stefan Kirschner**, **Gregor Heinrich** and **Lukas Langer** for the joint projects. A big tank you to **Ahmad Qais Parsa**, whom I could supervise during his apprenticeship. We both learned a lot during this time and I'm sure you will do well during the rest of your training.

No matter how good the supervision and the infrastructure is, without a nice atmosphere and great co-workers it won't be worth it. Thank you AK Bräse, especially Lab 306 for providing this essential positive mood and support. Thanks **Dr. Yuling Hu** for the second coffee of the day. Thanks **Janina Beck** for the voice of reason. Thanks **Dr. Alexander Braun** for the voice of outrage. Thanks **Dr. Florian Mohr** for crazy ideas in and outside of the lab. Thanks **Jasmin Busch** for dancing at PhD parties.

For the correction of this thesis I want to thank **Lukas Langer** and **Jasmin Busch**.

Finally, I want to thank my family. My mother and Alexis for the many ways of support and endless belief in me. You made me the person I am today.

For the last eight years I could always rely on one person and I enjoyed every day of it. Thank you Kathi for everything.